Scavengers of Beauty

A personal, cultural and symbolic
exploration of the Moon landing

Scavengers of Beauty

A personal, cultural and symbolic
exploration of the Moon landing

Philippe Sibaud

BOOKS

Winchester, UK
Washington, USA

JOHN HUNT PUBLISHING

First published by O-Books, 2020
O-Books is an imprint of John Hunt Publishing Ltd., 3 East St., Alresford,
Hampshire SO24 9EE, UK
office@jhpbooks.com
www.johnhuntpublishing.com
www.o-books.com

For distributor details and how to order please visit the 'Ordering' section on our website.

ISBN: 978 1 78904 474 4
978 1 78904 475 1 (ebook)
Library of Congress Control Number: 2019949610

Design: Stuart Davies

UK: Printed and bound by CPI Group (UK) Ltd, Croydon, CR0 4YY
US: Printed and bound by Thomson-Shore, 7300 West Joy Road, Dexter, MI 48130

We operate a distinctive and ethical publishing philosophy in
all areas of our business, from our global network of authors to
production and worldwide distribution.

Contents

Other books by the author

Poetry (in French)
Petale d'orage (1999) – Librairie-Galerie Racine, Paris
ISBN 2-243-03895-1
Tant pis pour le diable (2007) – Librairie-Galerie Racine, Paris
ISBN 2-243-04318-1
Dialogue avec l'ours (2010) – Editions Baudelaire, Lyon
ISBN 978-2-35508-409-6
Les Créates (2013) – Editions Baudelaire, Lyon
ISBN 979-10-203-0256-4

Fiction (in French)
Cannibalisme d'automne (2014) – Editions Baudelaire, Lyon
ISBN 979-10-203-0433-9

To and For Gaia

When you've cooked the marrow of the Sun and Moon,
The pearl is so bright you don't worry about poverty
*– **Sun Bu-e***

Introduction

Part 1

I

I reached Gare du Nord, Paris, in the dying light of a cool October day and jumped in a taxi to take me to the hotel. The year was 2006 and I was 42. The driver was Moroccan and very chatty. Once the usual niceties had been exchanged, however, the conversation took an unexpected turn. He was a sort of diviner, he told me, had a gift which sadly he was not using much, except when he was back home. I thought the whole thing was a bit unusual. A taxi driver pouring out very personal, and quite sensitive, material to a complete stranger? After all, diviners are not held in much esteem in our rational, no-nonsense world, and often get a frosty reception, to say the least. But I was lending him a sympathetic ear even though at the time I could in no way count myself as a diviner. Suddenly he stopped talking. A big black car – Mercedes type, I reckon – had overtaken us and I immediately noticed its somewhat odd licence plate – a triple 7 with some letters either side of it. Definitely not French registered. A diplomatic plate maybe? But I was not the only one to notice it. "Oh, Monsieur!!" he said. "Look at this car! I do not know why you are in Paris but you will go through great hardship!" A flash of inspiration, a sudden intuitive insight triggered by the sight of a car. This had escaped his lips before he had had time to think about it. He quickly realized his blunder. Diviner he may have been, but I sure had not asked him for his services. "But it will end up alright," he added. And I could not tell whether he had said it to make amends and comfort me or whether he had truly meant it.

II

This was my third day in police custody and I was mulling over the whole episode. I had come to Paris from London to be interrogated as a witness for a matter relating to my job as an oil trader with my former company, which I had left five years before. I thought it would take a few hours, maybe a day. As it turned out I had been placed under interrogation for two full days, spending two nights in jail. I was then waiting to be presented to the investigating magistrate for further grilling. The previous night had been difficult. I had spent it in a small smelly cell, crude lights relentlessly glaring all night long, shivering body with no blankets to hide under, and the wardens coming up every hour and playing with their keys in the door so as to disrupt the little sleep I could get. Kudos to the taxi driver, he had hit the nail right on the head. I could only hope that his parting words would prove as prophetic.

It was late in the day by now, a day I had spent in a glass cage waiting for the judge to audition me. "Softening the meat," they call it. I was crushed, dirty, dishevelled, humiliated, ashamed, desperate.

Suddenly I knew exactly what I had to do.

III

Fifty years ago, on 21 July 1969, a 5-year-old boy was stirred from his sleep by his parents to watch on a tiny black-and-white screen the blurred image of a man in white wearing a big helmet and awkwardly walking on dust. I have very faint memories of these images on the screen but the palpable sense of excitement that was then pervading the room has stayed with me ever since. "Philippe," said my father, "we are walking on the Moon!"

IV

Thirty-seven years separate these two events but as has become increasingly clear for me I cannot but see a striking parallel between them.

This is, however, only one part of the story, albeit the most dramatic one. Symbolically, I can see the whole narrative of my life shadowing the story of the Apollo mission. A kind of mythical resonance, reaching down to the same archetype. This figurative take, of course, is highly personal. It does not have any 'objectivity'. It is not 'true' and yet it is for me. And so, while I am mainly concerned in this book with the symbolism of the Moon landing, I will also explore this symmetrical unfolding, this twin pattern echoing through time and space.

What I am offering is not an academic work – yet it draws on numerous academic sources. It is not poetry – yet it is poetic. Nor is it a novel, even though it tells a story. Precariously it tries to tread the fine line between personal and collective, objective and subjective, literal and symbolic, astronomical and astrological, ultimately striving to reach the ambitious goal of reconciling the opposites, celebrating the union of masculine and feminine in the *hieros gamos* of the alchemists, *Sol y Luna* – the Sun and the Moon. In many respects this work is then nothing but a hybrid animal, a three-headed chimera, raging fire and moonlight dew, a Gorgonesque painting splattered with sacrificial blood, sound of blaring trumpets dipped in acid, smell of yellow fumes exhaled by dying panthers.

This book does not try to demonstrate anything, nor to prove a point. No Truth is to be found here. Or, if there ever was, it has long since splintered into the thousand eyes of a hallucinating spider. It is no reductionist postmodern flatland either. All it asks of the reader is a pinch of salt, ideally of the Himalayan pink variety: open-mindedness and an acknowledgement of the Mystery – whatever that means. And a willingness to hold paradox. *Especially* a willingness to hold paradox. A thistle is a bearded old man is a thistle. It will appeal to those who are not desperately searching for answers but who want the freedom to ask questions. To those who acknowledge the broken light leaking from Holy Scripture. Who can soar with the eagle and dig with the mole. Who can find Hope in the rotting flesh of a carcass, and the diamond sutra in a pile of dung.

It will appeal, in short, to the scavengers of Beauty.

Here is a tapestry, an Arachnean endeavour, weaving the golden rays of the Sun with the silver threads of the Moon, slowly revealing the ancient, radiant face of Earth. It tells the story of a homecoming, a space odyssey in the mirror of the soul.

Part 2

I

The tremendous effort that placed man on the Moon is worthy of admiration. By one reckoning, 400,000 people participated in this collective endeavour,[1] an incredible feat of coordination, dedication and scientific excellence. In 1970, the great mythologist Joseph Campbell immediately grasped the importance of what had happened, enthusing about a night when "our incredible human race... [had] just broke free of the earth to fly forth to the opening of the greatest adventure of the ages."[2] But beyond the heroic dimension of the Moon landing, is another reading possible?

"In that immense project... that sent man to the moon," said Jungian psychologist Edward Edinger,[3]

> it was Apollonian man, represented by the scientists and the planners and their ideas, who made that leap possible, while Hephaeistian man, signified by the engineers and the factory workers, made the equipment and the hardware that brought success. Arean man, represented by the astronauts, had the courage and the aggressive energy to make the trip, and Hermetic man, in those who are yet to come, will grasp the larger, hidden, and symbolic meaning of the arrival of man on the moon.[4]

Named after Hermes, messenger of the goddesses and gods, a deity held to be "the mediator of all hidden wisdom,"[5] the Hermetic man seeks to be a bridge, a messenger, a translator, an *interpreter*. "It is by *interpreting*," says French philosopher Paul Ricoeur, "that we can *hear* again."[6] This work, then, probes possible ways to interpret the *Apollo* mission – it proposes a *hermeneutics* of the Moon landing.

What is hermeneutics? The theory of interpretation is the short answer. But as historian of religion Jeffrey Kripal emphasizes, there is more than meets the eye: "in hermeneutics, as in quantum physics, there is a single process that co-creates both the subject and the object at the same time."[7] In other words, an interpretation is not a neutral exercise between an objective interpreter and a subject detached from him (her). Just as in quantum physics experience and experiencer are in a dance, a powerful feedback mechanism is at play in the hermeneutic undertaking – the 'hermeneutic loop', as Kripal calls it. One is interpreted as one interprets.

The act of hermeneutics is accomplished through the human imagination and its symbolic expressions. Its language is the image. The imagination, however, and crucially, is not to be regarded as a fanciful or innocuous movement of the mind. Quite the contrary, an important distinction must be made between flights of fantasy – the 'imaginary' – and the *imaginatio vera* of Paracelsus – the 'imaginal'. The imagination, in that latter sense, is a bridge, a bridge to the darkness within and without, a bridge, too, to radiant light; it is a medium that speaks to the unspeakable, to the Mystery, and in turn is spoken to – in images, often; in abstruse language, sometimes. A very serious endeavour it is – which does not preclude playfulness. With it, through it, we encounter *leela*, the divine play.

II

Let us try and close the hermeneutic loop. In my case: why bother

5

about researching so long and writing so extensively about the Moon landing? At the risk of being charged with navel-gazing, I have to turn the clock back on myself.

III

At this point I want to bring in the ancient practice of interpreting the Heavens, a.k.a. astrology, and I do it with much reservation and caution because I know that no other subject is as likely to potentially antagonize the reader.[8] There are probably as many definitions of astrology as there are astrologers. There are also different astrologies[9] but my focus in this work will be on natal astrology – the astrology concerned with birth charts – which I want to define as an act of imagination offering a hermeneutic of the psyche (for individual charts, e.g. myself) or an event (for collective charts, e.g. Moon landing). The chart can be compared to a mirror, in which one sees oneself expressed in astrological symbolism. Not because the chart (or the Heavens) has *caused* anything but because, in a way that nobody has ever been able to fathom, it *reflects* something.

And a mirror can be a very powerful tool. Perseus deflected the image of Medusa through his shield, held as a mirror. Only in this way could he confront her. Often the chart acts in a similar fashion:[10] we can only face ourselves, or help others face themselves, by holding the chart as a deflecting mirror, a decoy to avoid the stony gaze of the psyche. In this very process we gain some objectivity and detachment, while at the same time being immersed in subjectivity. We reach here at the heart of the "paradoxical practice of hermeneutics":[11]

As [the protagonist] reads and interprets the text of his life… he discovers that its story or plot changes. He discovers the circle or loop of hermeneutics. *He discovers that as he engages his cultural script as text creatively and critically he is rereading and rewriting himself.* He is changing the story (author's italics).[12]

This is crucial insight and Kripal has no equivalent in expressing it so vividly. As a hermeneutic practice astrology does not escape this loop, as Jung has spectacularly illustrated through his marriage experiment.[13]

What, then, do I see when I read my astrological script? Immediately jumping at me is the opposition between the Sun and the Moon. I was born eight hours away from a Full Moon, when both luminaries[14] are opposed. Strongly emphasizing this opposition is the presence of four planets conjoining the Sun (Mercury, Saturn, Mars and Chiron) directly opposing two planets joined with the Moon (Uranus and Pluto). In astrological parlance this is a classic 'see-saw' chart, when a group of planets are located 180 degrees away from another group. An opposition demands resolution, a meeting in the middle. Generally, each pole is visited in turn and a see-sawing takes place between extremities. In my own case, using broad symbolic brushstrokes, I feel that I have been erring on the side of the Sun for the first 42 years of my life, and on the side of the Moon ever since (twelve years and counting). The turning point took place during my three days in the Underworld of a French jail – my own transformative Moon landing. By exploring the symbolism of the *Apollo* mission, I am addressing a major archetypal feature of my life as reflected in my astrological chart. When I write about the Sun/Moon *coniunctio* of the mission, I am really writing about my own, elusive, *coniunctio*. And, in this very act, I am rewriting myself. Astrology, of course, only offers one level of interpretation. The ultimate level is, as ever, out of reach. But, as Whitley Strieber[15] says, when we face the darkness with an open mind, "the enigmatic presence of the human mind winks back from the dark."[16] And, one may add, other presences are felt too, grinning from the crannies in the walls of the mind.

IV

Here is a simple question, the very question that, notwithstanding its deceptive simplicity, set me on my quest:

Why was the mission to the Moon named after the God of the Sun? Would not a Moon Goddess (or a Moon God) have been much more appropriate if it was the Moon we were after?[17,18] It is not like there is any shortage of Moon Goddesses to choose from: Artemis, Hekate, Selene,[19] Ishtar, Isis, among many others. So why settle for a Sun-God? What does it say, if anything at all?

But first things, first. The name of the NASA programme that was to achieve President Kennedy's 1961 grand vision[20] was given by Abe Silverstein, Director of Space Flight Development at NASA:

> Abe Silverstein proposed the name 'Apollo' because it was the name of a god in ancient Greek mythology with attractive connotations and the precedent for naming manned spaceflight projects for mythical gods and heroes had been set with Mercury. Apollo was god of archery, prophecy, poetry and music, and most significantly he was god of the sun. In his horse-drawn golden chariot, Apollo pulled the sun in its course across the sky each day. NASA approved the name and publicly announced 'Project Apollo' at the July 28-29 [1960] conference.[21]

There is a delightful twist here – a tongue-in-cheek wink from the dark. 'Stein' in German means stone, and silver is the metal associated with the Moon. Decoded, 'Silverstein' stands therefore for 'Moonstone'. Abe, his first name, is short for Abraham, and Abraham is the father of the monotheistic, Abrahamic religions that will gradually sweep over the Western world from the fifth century BCE – solar religions all, as will be explored later on. So in the very name of the father of the *Apollo* programme to the Moon we find traces of the Sun/Moon polarity of the mission. The hermeneutic loop? Or am I making that up?

Apollo on the Moon. The Sun and the Moon. Unquestionably,

a multitude of images are conjured up by these two words in their fundamental tension, opening up a large vista of symbolic meaning. For, as symbolist Jules Cashford observes, "Moon and Sun fall into the pattern of creation as among the earliest beings to be created, sometimes before, sometimes after, Earth."[22] Beyond the accepted script of a heroic human conquest, can the iconic *Apollo* mission then be read through the symbolic interplay of Sun and Moon, and what can that reading teach us?

And, I may add, how can that *teaching read* us – read me?

V

I was standing on top of Taquile Island, on Lake Titicaca. Summer of 2009, three years after my traumatic experience with the French police. My life had taken a sharp turn, and it was about to take another significant inflexion. In Peru, as everywhere in the Andes, one encounters the powerful presence of Pachamama, Mother Earth, and many a traveller is changed by it. I was changed by it. I remember vividly experiencing the embrace of the Mother. She was all around me, and she was in me too. A Tarot reader had once told me that I had great love for Gaia. At the time I was surprised. I had never recognised, let alone acknowledged, this love. Worse still, I had spent 18 years in a job that had vastly contributed to pillage and desecrate the Mother. Or is there another way to look at it? A very special man, whom I can only describe as a kind of magus, had once remarked on my profession: "Oil trader? Ah! Working with the energies of the Earth!" That is indeed how I had always unconsciously felt about my job. I was a link in a vast stream of energy, an energy that had built up over millions of years and was suddenly released. I was in turn energised by it, part of the movement, an infinitesimal conduit, dancing with the Mother, spellbound. Until the shadow of the job hit home, that is.

I had stayed two hours with the diviner, and this off-the-cuff remark about my love for Gaia was the only thing that had stuck in my mind. Somehow it had resonated profoundly, struck a deep chord within, and its echo was now reaching me on this remote island, thousands of miles away from home, the dying light of the sun reflected on the glistening surface of a lake oh! so close to the stars. One year later I would be involved with the Gaia Foundation, a London-based NGO whose stated mission is to "uphold indigenous wisdom and an Earth-centred perspective,"[23] specifically through the ecological vision of Thomas Berry's Earth Jurisprudence. My Sun/Moon polarity had led me, by hook or by crook, to Earth.

Part 3

I

I want now to spend some time on the structure of this work, a work in two parts revolving around the actual Moon landing.

The first part deals with the voyage from Earth to Moon, and it is a voyage drenched in Apollonian consciousness – a lunar journey under the aegis of Apollo. Abe Silverstein was spot on: symbolically, Apollo was the perfect name for the lunar programme. There is indeed a very good reason why Mr Silverstein, a hard-core engineer at the vanguard of technological excellency, would be smitten with the Greek Sun-God: Apollo has come to symbolise the rational, scientific, objective view of Reality privileged by our Enlightened modernity. The Apollonian eye sees from a distance and is detached, dispassionate, cold, calculating, mechanical. It is also very good at what it does and, by Jove, how these engineers at NASA excelled! I have spent much time researching the unfolding events leading from Kennedy's 1961 very public pledge to the actual 1969 landing. One can only marvel at the technological prowess of the whole venture.[24] This, truly, is Apollo in its full glory – the Apollo, that is, who, in another hermeneutic loop, became a distortion of our own image. From that angle, the Moon landing can be seen as the pinnacle of the Enlightenment project, the ultimate conquest of the Sun over the Moon – in other terms, of solar consciousness over lunar consciousness. "We choose to go to the Moon in this decade..." said JFK. What wonderful words, the true offspring of our innate confidence in our abilities if we put our minds to it. "... and do the other things, not because they are easy, but because they are hard!"

The Moon, which had entranced humanity for millennia, has been the chief causality of this Apollonian takeover. She has been gradually stripped of her veils, one by one, until, in the summer evening breeze of a dying decade, she stood stark naked and lifeless. We have walked on her body, a glistening carcass aimlessly wandering in the dark of space, a fallen deity with potholes on her face. "A mythical world becomes real estate," exclaimed sci-fi writer Arthur C. Clarke.[25] Defeated by the stinging rays of a merciless Apollonian consciousness, the Moon's epitaph reads:

Born at the dawn of humanity
Died July 20, 1969

II

I suppose I needed it badly. Not that I was particularly arrogant or full of myself (I hope). But success comes with a price: a certain hubristic feeling of invulnerability and – this is harder to admit – a diffuse sense of entitlement. But take my word: being stripped naked, asked to bend over and having your most private part inspected lest you hide anything there – my own intimate Moon exploration – is a very effective antidote against self-inflation.

But one thing is clear: the man who left that police station on Thursday evening was aeons away from the one who entered it on Tuesday morning.

III

Something happened on the surface. Something happened deep down too. A Moonquake, a convulsion. The astronauts who went there were not the same when they came back. "We went to the Moon as technicians, we returned as humanitarians," said Apollo 14's Edgar Mitchell.[26] We, too, are no longer the same. Maybe this should come as no surprise. By reaching the outer Moon, how could we expect the inner Moon to remain unaffected? Besides, the Moon had always stood as a symbol of transformation. "When the Sun is the invincible conqueror of death," says Cashford,[27] "the Moon surrenders to death as to the ambivalence of life, yet lives to rise again."

The Moon cycle is constant change within eternal time – death and rebirth, again, and again, in a never-ending cosmic dance. And when the Sun-God lands on the Sea of Tranquillity, we may be entitled to ask: does Apollo kill the Moon, or does he kiss Her? By shifting our perspective, we may suddenly see the Sun conjoining with the Moon, his Heavenly Bride. A New Moon, when a new cycle is initiated. Rebirth. No longer conquest, but union. Or communion. Buzz Aldrin, the second man on the Moon and a devout Christian, took with him a wafer and wine and quietly performed communion on the Moon. Is this to be seen as the aggressive act of a conquering religion, much like the planting of the American flag, a thorn in the lunar flesh? Or is something else symbolically at play? Can the Moon landing be seen as the alchemical sacred marriage *Sol y Luna* – a marriage, what's more, giving birth to the *infans solaris lunaris*, Earth, the offspring of Sun and Moon?

By taking leave, mankind has for the first time been able to view our planet from afar, and that view, symbolised by the iconic *Earthrise* picture taken by Apollo 8, has had 'earth-shattering' consequences. Echoing similar developments taking place in the 1960s, a new consciousness arose.

By the operation of the Moon landing, Earth, the blue marble, the lapis lazuli, has seemingly been reborn, cloaked in an aura of intense numinosity. She has become spiritualized again, a divine child delicately poised in the fecundating darkness of space. "This Earth," said Joseph Campbell, "an extraordinary kind of sacred grove, as it were, set apart for the rituals of life... the entire globe now a sanctuary, a set-apart Blessed Place."[28] In that sense the *Apollo* landing, seen as a Sun/Moon *coniunctio*, may provide a mythopoeic foundation for the new narrative emerging in the 1960s: the emergence of a planetary sensibility, a Gaian consciousness, seeping in the collective through the Gaia archetype.[29]

A new gaze is dawning, reminiscent of the magnificent last scene of Kubrick's 1968 *2001: A Space Odyssey*, in which a wide-eyed foetus floats above a translucent Earth. No wonder conspiracy theorists suspected the same Kubrick to have been commissioned by NASA to fake the Moon landing in a well-hidden cinema studio. As a true artist he was the harbinger of a new consciousness, tapping into the collective before (almost) anybody else, and such artists always come with a whiff of suspicion.[30]

IV

In 1967, the wind cried Mary.[31] In 1969, a whisper could be heard on the lunar surface – heard, that is, by those not deafened by the sound of exploding rocket fuels – a whisper that cried Earth.

This cry is the object of my second part, the return journey – an earthbound journey under the aegis of Gaia, our very beautiful and very sick Mother.

Part I

From Earth To Moon

The rising rocket appeals to instincts older than reason; the gulf it bridges is not only that between world and world – but the deeper chasm between heart and brain.
– Arthur C. Clarke

Can you hear the drums
Rolling
Over the land
Of milk and honey?

Can you see
Tiger
Dreaming the dream
Of butterfly?

Let us play strip poker!

And,
In the twilight of the Gods,
May we be the Naked Losers,
The Holy Fools!

Chapter 1

Moon and Sun

The Evolution from Lunar to Solar Consciousness

I am the Sun, the Moon, and the many piglets suckling on the great breast of the Milky Way.

I have always been totally flummoxed by the fact that Sun and Moon appear of similar size in the sky. How extraordinary. Their respective rulership over day and night is equally fascinating. What an incredible polarity, the stuff of magic. And yet this is readily dismissed by science as coincidence, with a kind of so-what attitude that has the unnerving effect of pulling the carpet from under your very feet, leaving you feeling somewhat stupid to have marvelled at this astonishing fact. I often wonder how humanity would have developed without Sun and Moon. Not literally, obviously – we know the answer to that – but symbolically. One indeed could not have devised better hooks on which to hang the number 2, the number of polarity. Or was polarity actually born of the very interplay of Sun and Moon? If we had had a third object in the sky – a second Moon maybe, with different rhythms – would consciousness have evolved differently? For as Edinger, following Jung, has often pointed out, consciousness arose out of polarity. The One is undifferentiated, the Uroboros, containing all opposites, a Garden of Eden of unconsciousness. With the Fall, as in many creation stories all over the world before and after, consciousness made its entrance on to the stage in dramatic fashion. Adam and Eve are naked and the rest is history, a history of polarity – yin/yang, Shiva/ Shakti, Ch'ien/K'un, Apollo/Dionysus, Sulphur/Salt, Sol/Luna. The story of humanity is the story of a chase, of a remembrance, a longing, a striving for the Sacred Marriage, the ecstatic reunion of the opposites, a quest for the Divine Consort. This is the fabled *hieros gamos* of the alchemists, which Jung used as psychological image for the successful integration in the psyche of feminine and masculine elements. It is also found in the image of the androgyne. "In his discussion of Hindu Tantric literature," says Jungian June Singer, "Eliade stresses that 'androgynisation' is only one part of a total process, that of the reunion of opposites. He speaks of 'opposite pairs' that have to be reunited: the Sun and the Moon have to be made one."[1]

Truly, one could not have designed a better pair than Sun and Moon. This is almost too good to be true. I find the Full Moon particularly moving, when the Sun sets at one end while simultaneously the Moon in her full glory rises at the other – with us, Earthlings, delicately poised in the middle, holding both realities at once, arms outstretched, one eye orange and one eye white. Charles Trenet, a popular French singer affectively nicknamed the Singing Fool, once sang that the Sun had a meeting with the Moon but that the Moon never made it. These two are never seen together, he wrote. It is a beautiful song, very catchy, crazy in the way that only Charles Trenet could be, who was born under a gibbous moon, less than two days from a Full Moon. At that point in the cycle the Moon is indeed chasing the Sun, about to reach the climax of separation in the Full Moon before eventually conjoining in the New Moon two weeks or so later. Novalis, the German Romantic poet, was born under this latter phase, just before the New Moon, a time of darkness when the Sacred Union of Sun and Moon is about to birth a new cycle. And birth Novalis certainly did, although his residence on Earth was cut short at the tender age of 29, the age of the Saturn Return,[2] a time of reckoning.

I had always assumed a sort of parity between the two 'stars',[3] Sun and Moon. One for the day, one for the night, a harmonious tick-tock, balancing each other out, a well-crafted celestial dance. And then one day – twenty years ago or so – I read Joseph Campbell's *Occidental Mythology* and things started to take on a much deeper hue. At some point in human history, says Campbell,

the celestial orb to which the monarch is... likened is no longer the silvery moon, which dies and is resurrected and is light yet also dark, but the golden sun, the blaze of which is eternal and before which shadows, demons, enemies, and ambiguities take flight. The new age of the Sun God has dawned, and there is to follow an extremely interesting, mythologically confusing development (known as *solarization*), whereby the entire symbolic system of the earlier age is to be reversed, with the moon and the lunar bull assigned to the mythic sphere of the female, and the lion, the solar principle, to the male.[4]

A *solarization* process... I was fascinated, a traveller suddenly discovering a new landscape, a rich place teeming with life and fabled creatures. There was obviously more to the Sun/Moon story than I had ever imagined. I had never ventured beyond the gate of their physical polarity, never noticed the symbolic perfume in the air. Campbell tore the curtain and I eagerly moved across the threshold. The story of consciousness, I discovered, could be seen through the angle of the Sun/Moon interplay. This spoke to me at a deep level and today, reflecting on it, I cannot but not bring it all back to my own exacerbated Sun/Moon astrological polarity. The hermeneutic loop. One is read as one reads. "We are not the artists but the drawings,"[5] said Philip K. Dick. Indeed.

If we peep through the lens of Sun/Moon symbolism, then, what do we see? How does our human story unfold?

First, a word of caution, a health warning. Symbols are tricksters, shape-shifters, notoriously resistant to being cornered and caged. To place them in a straightjacket of meaning not only devitalizes them but is the quickest road to literalism – the murder of the symbol. A light touch is required. In Plato's words,[6] we are contemplating here a 'likely story', not absolute truth. A mythical take. A gaze from a particular vantage point. As Marsilio Ficino magisterially expressed in the opening sequence of the *Book of the Sun* – and I make mine his words:

> Most careful reader, be indulgent to me – just be mindful of the Apollonian and as it were poetic licence before the Sun, while not disallowing me a more serious (and as the Greeks say) dogmatic content. I have promised an allegorical, and to that extent, a mystical exercise of the wits, in the name of Phoebus the oath-orderer, whose gifts these are.[7]

With poetic licence, then, let us try and imagine what it must have felt like for a Palaeolithic hunter-gatherer. By day the Sun moves, regular, unchanging, reliable and therefore unnoticeable. But night-time tells a very different story. There, in the midst of countless sparkly points, is this enigmatic orb, never the same yet following a regular pattern of appearance and disappearance. How utterly fascinating. How to account for it? And how not to imbue it with very special qualities? How not to stand in awe? Here was a living entity and it must have been hard not to see in it a reflection of earthly life and death. In so investigating the psyche of a hunter-gatherer I want to stay alert to my own prejudice as a 21st century Westerner. I have no doubt that the level of sophistication was much greater in the Palaeolithic than is ever being acknowledged – Lascaux and Altamira provide ample evidence of this[8] – and I want to do justice to our seminal ancestor. I certainly do not want to patronize or caricature him, or cloth her with my own projections. As far as we know, though, the first works of art of that period were connected to the Moon and to the Feminine principle – the Venus of Laussel, the Goddess of Willendorf, and countless others[9] – and it is hard not to conclude therefore that primacy was given to the Moon, connected as She seemed to be to the cycles of Life observed on Earth.

An important point has been made here – namely the association between the Moon, the Earth and the Feminine. This association will indeed inform my work but with an important caveat: Feminine must not be confused with woman, nor Masculine with man. In other words, gender and archetype should not be conflated. This controversial issue has infused much of the work done by feminists and non-feminists alike, with a tremendous amount of confusion in-between, and I want to stay clear of the fray. Rather, I will align myself with Jules Cashford, when, following Jung and Neumann, she claims that

> the Moon is a vessel, cup or chalice… which belongs to the night, was constantly related to water…, was identified with the Earth and the Great Mother Goddess, and was the presiding force (whether male or female) in the earlier ages of matriarchal consciousness.[10]

In this view, the Moon is archetypally feminine and the Sun archetypally masculine, even though they have been mythologized through gods and goddesses alike, depending on the cultural context.

As Cashford notes, the many figurines and statues depicting Mother Goddesses, usually replete with lunar symbolism, seem to point to an earlier age of matriarchal consciousness. Some researchers go even further – such as Chris Knight, a social anthropologist steeped in Marxism whose work I haphazardly stumbled upon one rainy October day. In a highly original work, *Blood Relations*, Knight develops the idea that culture originated with women through menstruation. The argument goes like this: in order to ensure maximum sexual fertility and simultaneously make sure that the men would go hunting and feed the tribe, the women would gradually synchronize their periods and go on sex strike until the men came back with meat. A simple equation: no meat, no sex. The most convenient tool to measure time was the Moon, says Knight, and thence developed a pattern whereby women would bleed at the New Moon, at which point they would become sexually unavailable. Men would then go on the hunt, not to come back before the Full Moon when women would reach the highest point in their fertility cycle. Sex and meat would then be on the menu. From this purely utilitarian construct, culture would develop, as Knight explains:

> With this model in place, I saw most aspects of kinship, ritual and mythology in traditional cultures as expressive of its logic. Most hunter-gatherer cultures... *did* sustain menstrual avoidance of one kind or another, *did* see prior sexual abstinence as essential to hunting luck, *did* link such abstinence with menstrual avoidances, *did* construct mythological connections between the blood of women and the blood of game animals, ... – and so on.[11]

This theory has been strongly challenged and I am in no way qualified to discuss its anthropological merits. No doubt Knight's strong Marxist background informed his view: women going *on strike*? I see Hermes grinning in the background. But I like it, I find it very alive and interesting. Of course, it is hard not to detect a typical male prejudice in this take. Namely, seeing women as Manipulators – although Knight no doubt would vehemently refute the accusation. After reading Knight's book I attended a lecture of his in London. Vaguely expecting him to linger on his pet subject, I was surprised to discover that, instead of serving us sex and meat, he proceeded to explore the *Sleeping Beauty* fairy tale. Not that it was uninteresting, quite the opposite. Away from Walt Disney's sanitised version, Knight emphasized the symbolic dimension of the tale: *Sleeping Beauty*, said Knight, is a *solarization* myth. It expresses the takeover of solar consciousness, as expressed by the respective roles of numbers 12 and 13 in the tale.[12] Only twelve fairies had been invited to the christening of the new princess. The thirteenth fairy – the bad fairy, which psychologist Bruno Bettelheim symbolically linked to the fatal 'curse', menstruation – had been left out and took revenge by cursing the poor child – and the rest is the stuff of dreams. We encounter here the classic demonization of the number 13, the lunar number (13 Moons in a 12-month solar year) and the imposition of the new solar order based around the number 12. *Sleeping Beauty*, then, expresses in symbolic form a celestial battleground: Sun versus Moon. Or rather, solar consciousness versus lunar consciousness, as Cashford illustrates at length:

It only needs the simple word 'metaphorically' to turn the stories of Sun and Moon into symbols of human consciousness... For 'daylight' consciousness belongs to the Sun as 'night-time' consciousness belongs to the Moon, and each represents to the human mind a different kind of illumination of itself... Further, Sun and Moon appear to move each against a different background, one of light and one of dark. It makes sense that they were given separate realities to inhabit, and that the terms 'solar' and 'lunar' consciousness are even now employed to represent different modes of thinking or modes of consciousness, though the meaning of these terms has moved through as many mutations as the meanings of Sun and Moon themselves. The current meaning at any point in history cannot then be 'natural' – inherent in the natures of Sun and Moon – but must be cultural, or usually, culture calling itself natural.[13]

The last point is important. It dispenses one from seeing eternal truths shining from within the symbol. A symbol is only a symbol for the psyche observing it, as Jung remarked.

Within the framework of our own culture we may therefore start to discern behind Campbell's *solarization* process the contours of a grander arc of duality: Sun/Moon, solar/lunar consciousness, Reason/Intuition, Intellect/Feeling, Masculine/Feminine, Patriarchy/Matriarchy, Conscious/Unconscious, Light/Dark. These well-known dualities operate differently within a solar or a lunar context. As Campbell pointed out, a solar consciousness is by its very nature polarized. The Sun does not accommodate the Shadow, Light is the enemy of Dark – an either/or perspective. By contrast lunar consciousness is much more fluid, for in the Moon light and dark not only cohabit but actually need each other – a both/and perspective. As Jules Cashford emphasized above, both modes respond to different human modes of experiencing reality, and both must be acknowledged for human consciousness to be complete. The Sun conquers but the Moon transforms.

In the beginning, then, was the Moon… but after reaching the fullness of her cycle sometime in the Neolithic, the power of the Moon started to wane and a new order centred around the Sun emerged.

Marija Gimbutas, in her groundbreaking work on the civilization of Old Europe, offered historical reasons for this shift: the arrival, in three major waves, of Indo-European invaders from the East – principally the Kurgan people from the Volga – who, thanks to the domestication of the horse and a war-oriented mentality, overpowered the peaceful, matriarchal indigenous cultures of Europe. The Kurgan were operating along a very different cosmology, she said, and brought with them a strongly patriarchal flavour. The Sky-Gods replaced the Earth Goddesses. "During and after this period," she wrote, "the female deities, or more accurately the Goddess Creatrix in her many aspects, were largely replaced by the predominantly male divinities of the Indo-Europeans."[14]

Erich Neumann, while reaching similar conclusions, came from a very different angle, in which phylogeny meets psychology. A true heir to Jung, Neumann dogmatically equated Sun with Masculine, Light and Consciousness, and Moon with Feminine, Darkness and the Unconscious. Too dogmatically, one might say, which would lead James Hillman to rather unmercifully denounce the "absurdities of Neumann."[15] There is, however, much to like in Neumann's work. His main premise is that the takeover of the Sun from the Moon was a necessary stage in human evolution. Just as a child grows into an adult by gradually separating from the embrace of the mother and by developing an ego, humankind had to emerge from the unconscious embrace of the Great Mother, with her lunar attributes, and develop a separate ego, symbolised by the Sun. Joseph Campbell concurred, identifying four successive stages in this gradual process:

1. The world born of a goddess without consort
2. The world born of a goddess fecundated by a consort
3. The world fashioned from the body of a goddess by a male warrior god
4. The world created by the unaided power of a male god alone.[16]

The first two stages correspond to Great Mother mythologies and run from the dawn of humanity to the late Bronze Age and early Iron Age (c. 2000 BCE). The last two mirror the appearance of male deities, culminating in the Father figures of the monotheistic, Abrahamic religions.

Ginette Paris, a Jungian and archetypal psychologist, offers a view elegantly blending Gimbutas with Neumann:

These patriarchal divinities, thirty-five hundred years ago, presented the 'new values' of the epoch... the new legislative power, represented by Zeus, and the rational clarities, personified in the classical period by Apollo... and one may guess that they were implanted, not only by force but also because they offered an escape from the hegemony of the Mother Goddess, perceived by some of her own children as restrictive and retrogressive.[17]

One has of course to resist the temptation of being too literal about it. This arc of development is painted in very broad lines. Campbell has been repeatedly challenged on his somewhat monolithic view. The archetypes of Great Mother and Great Father must be handled with utmost care, at the risk of tripping feet first on to a reductive and not very helpful nor subtle approach to the incredibly rich and multilayered story of humanity. They may also highlight our obsession with monotheism, as emphasized by Ginette Paris:

If we can let go of the devotion to an original, single matriarchy of the Great Mother (that idea which supports monotheistic feminism), then we can regard the plurality of goddesses, not as her fragmentation or as her developmental differentiation, but rather as each goddess comprising an archetypal form of feminism.[18]

As for me, at the risk of simplification, I cannot but not be seduced by a very aesthetic line that runs like this: the Moon was the main celestial object in the sky for millennia before being gradually superseded by the Sun. Behind the Moon we find the archetypal Feminine, and behind the Sun the archetypal Masculine. Solar consciousness took over in the Iron Age and has not relinquished power ever since.

And maybe this was a necessary development, as Neumann asserts. But it came at a very heavy price – the repression of the Feminine. "For every step and development that presses toward patriarchal consciousness – that is, toward the sun," says Neumann,[19] "the moon-spirit becomes the spirit of regression, the spirit of the Terrible Mother and the Witch." The nascent ego, says Neumann, terrified of losing its fragile autonomy, will ferociously do whatever is necessary to preserve it. The fear of the Feminine is this absolute terror, the fear of castration, the fear of being engulfed back into the unconscious, the fear of losing one's identity. The Feminine in her dark aspect – Kali, Lilith, Ereshkigal – is absolutely petrifying for the Masculine. But even – or maybe especially – in her destroying aspect the Feminine can be said to be in service to Life. She destroys to regenerate, she kills to give birth. This is the Moon, which has to die to be reborn – a totally unacceptable sacrifice to a solar ego that refuses death at all cost. The Masculine, in that sense, is not in service to Life but to life – his own. But only at this price – the price of separation – is consciousness able to emerge. One has to sever the connection from the All Encompassing Unconscious, or one is doomed to err forever in a state of unrealization, like a child forever stunted by his mother and unable to reach a state of psychological autonomy. Ken Wilber followed a similar model of development in his own work. But, he stressed, our 'overidentification' with the ego has brought us to a dead end. We have been overpowered by the Sun's light. Apollo rules and Artemis is exiled.

In writing the lines above I am aware of the very fraught nature of this debate. Male/Female, Masculine/Feminine, Sun/Moon… these notions have become more than ever contentious and blurred. But one has to hang one's hat somewhere. I am an unapologetic essentialist. I like the dance of polarity, I find it beautiful and rich and immensely creative. Blame my astrological make-up if you will…

The mention of Apollo at this juncture is not without ground for, as Campbell emphasized, the *solarization* process was accompanied by the emergence of many new solar myths. Greece was the main battleground for this clash of soli-lunar consciousness – a fertile ground that has graced us with numerous myths of invaluable insight and poetry – but the rupture point, between Campbell's stages 2 and 3, may be said to have happened around 2000 BCE further East with the *Enuma Elish*, the Babylonian epic of Creation, in which the sun-god Marduk overwhelms and kills his great-great-great-grandmother Tiamat. This is done in pretty gruesome fashion: "Marduk treads on her legs, crushes her skull, severs her arteries, and splits her like a shellfish into two parts. Half of her he makes into the sky and half of her into the Earth."[20] Following Neumann, we can identify here two of the main elements of these new myths:

1. The appearance of a solar hero – the ego – killing a monster, usually of a feminine nature – the forces of the unconscious.
2. The separation of the World Parents, Heaven and Earth, by the hero, representing the birth of consciousness.

Furthermore, say Baring and Cashford,

in the *Enuma Elish* there is already the germ of three principal ideas that were to inform the age to come: the supremacy of the father god over the mother goddess; the paradigm of opposition implicit in the deathly struggle between god and goddess; and the association of light, order and good with the god, and of darkness, chaos and evil with the goddess.[21]

The stage is set for the new world order, and countless myths of Creation will replicate this blueprint. In Greece one such myth is the castration of Ouranos by his offspring Kronos bringing about the separation of Heaven (Ouranos) and Earth (Gaia), the original parents. But while Gaia, the Mother Goddess, is in this particular story rescued by her child from the clutches of an overbearing husband, many other myths tell of the victory of a masculine solar hero over a telluric feminine monster – Perseus and Medusa, Apollo and the Python, Bellerophon and Chimera, Herakles and Hydra, to name some of the most famous Greek ones. Scholar Jane Harrison would call this male murder frenzy a "protest against the worship of Earth and the daimones of the fertility of Earth."[22]

Not everyone, however, agrees with this psychological interpretation of myths. For such a towering figure as Robert Graves, for instance, "a large part of Greek myths is politico-religious history" – recording the historical changes that were happening at the times of the Aryan invasions: "[Perseus] was not, as Professor Kerényi has suggested, an archetypal Death-figure but, probably, represented the patriarchal Hellenes who invaded Greece and Asia Minor early in the second millennium BCE and challenged the power of the Triple-goddess."[23] And further: "Again, Apollo's destruction of the Python at Delphi seems to record the Aechaens' capture of the Cretan Earth-goddess's shrine."[24] This more literal interpretation of myths was strongly challenged by the ever-symbolist Campbell but these two takes may in fact not necessarily be incompatible with each other, as Ginette Paris illustrated earlier.

Nowadays, myths tend to be discounted as old-fashioned imaginary stories – "quaint relics of the nursery age of mankind," in Robert Graves' words[25] – at best poetic constructions, at worst harmful delusions, and this attitude to myths is in fact the worst delusion of all. For indeed myths are actually reflections of a particular consciousness, and this consciousness is visibly expressed in the social order of the day. Before the arrival of the invaders from the East, says Campbell,

> there had prevailed in [the ancient world] an essentially organic, vegetal, non-heroic view of the nature and necessities

of life that was completely repugnant to those lion hearts for whom not the patient toil of earth but the battle spear and its plunder were the source of both wealth and joy. In the older mother myths and rites the light and darker aspects of the mixed thing that is life had been honored equally and together, whereas in the latter, male-oriented, patriarchal myths, all that is good and noble was attributed to the new, heroic master gods, leaving to the native nature powers the character only of darkness – to which, also, a negative moral judgment now was added.[26]

We are the inheritors of this *Weltanschauung*, this worldview, true heirs of a distorted vision that cries for a new story.

Like no doubt many people, I have always felt a close affinity to myths – without really reflecting on the why and the how. When I was a child – around the age of 10 – I was absolutely entranced by the *Iliad* and the *Odyssey*. My parents had offered me trimmed down, child-friendly versions of the Homeric epics and I devoured them repeatedly. I grew up in a French town called Troyes, a word that is phonetically similar to the *Iliad*'s Troy. Is this the reason why I always vouched for the Trojans? For me the Greeks were a bit too self-righteous and bloodthirsty. I could not help but feel tremendous sympathy for the Trojans, despite the pain of finding myself on the losing side at each and every rereading. Besides I always thought that the true hero of the *Iliad* was Hector. I had no time for Achilles and his sulking ways. It was too easy for him anyway, what with that quasi-invulnerability and all that jazz, he did not really seem human. But Hector was another kettle of fish, a gem of a man who exuded real depth and humanity. I have never been able to forgive Achilles for dragging Hector's corpse like a piece of meat around the ramparts of Troy. This left a very black spot on his character. Had he been victorious Hector would never have displayed such cruelty – a sign of unconsciousness, really, a lack of compassion. Hector in the *Iliad* and Ulysses in the *Odyssey* were my companions, male ideals for the developing boy that I was. Archetypal figures, both very masculine and yet not estranged from their inner Feminine – as I realized much later. Displaying an interesting polarity between them, too. Hector is the quintessential family man, a man, too, of his land and of his people, rooted in the collective values of his tribe, a beacon, a pole star for others to navigate by. He is Apollonian in the most positive aspect of the word, shining forth with clarity, the Apollo of Delphi, of 'Know Thyself' and 'Nothing In Excess'. Ulysses, although also longing for home, wife and land, is a very different figure, much more of a loner, not afraid to bend the rules, at times tricksterish, a Hermes character. In him the individual always seems to take precedence over the collective – a counterpoint to Hector's moral stance, who walks to his death with open eyes out of sheer duty.

"The patron god of the *Iliad* is Apollo," says Campbell,

the god of the light world and of the excellence of heroes. Death, on the plane of vision of that work, is the end; there is nothing awesome, wondrous, or of power beyond the veil of death, but only twittering, helpless shades... the end of it all is ashes. In the *Odyssey*, on the other hand, the patron god of Odysseus' voyage is the trickster Hermes, guide of souls to the underworld, the patron, also, of rebirth and lord of the knowledges beyond death.[27]

Can the *Iliad* be seen as a solar journey, then, and the *Odyssey* as a lunar one? I only recently came across this quote by Campbell while working on the Moon landing but it made a deep impression on me. I strongly felt that it mirrored the *Apollo 11* mission: a solar journey of conquest to the Moon under the aegis of Apollo, a transformative return journey back to Earth under the aegis of Hermes. I sensed an interesting parallel and I adopted it for my work, although in the end I placed the return journey under the aegis of Gaia rather than Hermes – who looms large in my work, however.

That the *Iliad* is a solar journey can be evidently seen in the figure of Achilles. Achilles' choice is to die young in glory, rather than old in what would seem (to him) an uneventful life. By the extent of his deeds his glory will survive him and his name be remembered long after his death. Here is a typical solar hero, terrified of death, who strives for immortality through his name – and by being an agent of death. By contrast Odysseus' journey back to Ithaca is very different. In the *Odyssey* Ulysses only kills twice. The first time just after leaving Troy when he lands with his men in the country of the Cicones and where, still drenched in the bloody consciousness of the *Iliad*, he pillages and murders. Note, however, that he then has to escape with his men when they eventually find themselves overpowered and at risk of death. At this point the pattern is set. Ulysses will spend the rest of the return journey *escaping*: from the Lotophagi, from Circe, from the Sirens, from Scylla and Charybdis, from Calypso. This is a very interesting development. Marie-Louise von Franz has highlighted in her work[28] how the path of transformation for women was through confronting and escaping – the escape itself being achieved through the integration of an important psychological lesson. As Campbell has shown in *The Hero with a Thousand Faces*, the masculine hero, by contrast, is transformed through conquest – the murder of some archetypal force represented by a monster. In that sense Ulysses' transformation in the *Odyssey* is very feminine: no solar conquest but lunar death and rebirth. Even the Cyclops Polyphemus, a proper chthonic man-eating monster, is not killed but merely blinded. Primarily though, Ulysses transforms through his encounters with the Feminine, especially Circe, Calypso and Nausicaa. But his true consort is Penelope with whom, says Campbell, he forms the archetypal Sun/Moon couple. Ulysses has an affinity with rams, a solar animal – it is indeed by hiding under the thick coat of rams that he and his men can escape from the Cyclops' cave. Penelope, by all intents and purposes a prisoner in her own home, is weaving by day and unweaving by night, a desperate measure to keep the suitors at bay and a movement akin to the wavering moon. The return home is the stage for the second act of killing by Ulysses, when he rids his palace of the unpalatable contenders – but the consciousness brought to that killing is very different from the consciousness of the pillaging of the Cicones. Ulysses has turned into a God:

> Then fierce the hero o'er the threshold strode;
> Stript of his rags, he blaz'd out like a God.
> Full in their face the lifted bow he bore.
> And quiver'd deaths a formidable store;
> Before his feet the rattling show'r he threw,
> And thus terrific to the suitor crew.[29]

When Ulysses returns home, then, the sacred marriage is taking place. Sun and Moon unite. Campbell notes how Ulysses reaches Ithaca in the twentieth year following his departure, a number that calls to mind the Metonic cycle. Meton, a Greek astronomer of the fifth century BCE, calculated that 19 solar years equated to 235 synodic lunar months. Every 19 years the phases of the moon recur on the almost exact same days of the year – a cycle that may or may not have been known to Homer, who was writing a mere 300 years or so before Meton. Robert Graves calls 19 the 'golden number' because it reconciles solar and lunar time, a fact that he surmised must have been known to our Neolithic ancestors in Britain who were often building their circles with 19 stones, notably at Stonehenge: "the number 19 is commemorated at Stonehenge in nineteen socket-holes arranged in a semi-circle on the Southeast of the arched circle."[30] Maybe Meton got all the plaudits by merely articulating the theory better than anybody else. Quoting Professor Gilbert Murray, a scholar of Antiquity, Campbell affirms that Ulysses rejoined his wife at the beginning of the twentieth year, that is as soon as the nineteenth year is completed: "on the last day of the nineteenth year, which is also by Greek reckoning the first of the twentieth, the New Moon would coincide with the New Sun of the Winter Solstice; this was called the meeting of Sun and Moon."[31] I do not want to make too much of it but I find the poetry absolutely irresistible.

The *Iliad* and the *Odyssey* may stand as both the first and the last of their kinds, marking an important transition – and maybe this is where their compelling power lies. While often recognised as the first works of Western literature they may also be the last remnants of a magical mentality in which Gods, Goddesses and Men (much less so women, however) constantly mingle and interact. For psychologist Julian Jaynes, Homeric consciousness was very different from our own:

> Iliadic man did not have subjectivity as do we; he had no awareness of his awareness of the world, no internal mind-space to introspect upon. In distinction to our own subjective conscious minds, we can call the mentality of the Myceneans a *bicameral mind*. Volition, planning, initiative is organized with no consciousness whatever and then 'told' to the individual in his familiar language, sometimes with the visual aura of a familiar friend or authority figures or 'god' or sometimes as a voice alone. The individual obeyed these hallucinated voices because he could not 'see' what to do himself.[32]

In other words, gods were voices from a hidden chamber of the mind. This is a fascinating angle and it seems to stand midway between Levy-Bruhl's *participation mystique* and Socrates' *daemon*.

After Homer, however, men took an increasingly independent stance. The gods receded in the distance, further and further away, in their Heavenly abode. "A real difference of religious outlook separates Homer's world even from that of Sophocles, who has been called the most Homeric of poets,"[33] said scholar of Antiquity ER Dodds. Dodds saw a growing anxiety taking hold of the Greek mind, which he variously attributed to challenging social conditions and to a loss of familial piety. But if Dodds is right, could the increasing sense of dread of Greek Man, ever more deserted by his gods and goddesses, be a harbinger of our modern and postmodern existential angst? For by the early 20th century God, famously, was dead and nowhere to be found, even where he was supposed to dwell: "that their spacecraft had encountered neither God himself nor any of his angels was asserted by Soviet scientists and cosmonauts to be evidence that such supernatural beings did not exist."[34] The solar Hero, increasingly human, has chased the divinities away and the world is a different place:

"Patriarchy," said Jungian Marion Woodman,

originated in one of the oldest myths of humankind: the hero's journey. In this myth, the hero is the descendent of the sun god, that symbol of absolute authority upon which all life depends. The sun god continuously reasserts his absolute authority by the conquest of the forces of darkness that challenges his reign... One symbol of this darkness is the moon with its lunar cycle. The cycle [is] essentially feminine... The feminine, standing for the forces of darkness and chaos, is brought within the orbit of a masculine light-bringing creation as a reflection of its power.[35]

Chapter 2

The Primacy of the Sun

The Sun in the Last Two Thousand Years of Western Civilization

And God made two great lights: the greater light to rule the day, and the lesser light to rule the night.
– Genesis 1:16

Plato's ambivalent attitude to myths is well documented: "Then came the early Greek philosophers," said Robert Graves, "who were strongly opposed to magical poetry as threatening their new religion of logic… elaborated in honour of their patron Apollo."[1] For Graves, this estrangement from mythical language was at heart an estrangement from the Goddess: "Socrates, in turning his back on poetic myths, was really turning his back on the Moon-goddess who inspired them and who demanded that man should pay woman spiritual and sexual homage."[2] By the time of Socrates (5th c. BCE) and Plato (5th/4th c. BCE), the Sun ruled in the Heavens. In the *Republic*, Plato compared the Sun with the Good and then proceeded with arguably his most powerful myth, the myth of the cave, in which a man escaped from the misleading shadows of his prison-like cave and reached the light of the Sun – metaphorically the source of the True, the Good and the Beautiful:

> The way things appear to me is that in the sphere of the known what is seen last, and barely seen, is the form of the good, but that when it *is* seen, there can be only one conclusion – that it, in fact, is cause for everything of everything right and beautiful, as both progenitor of light and of the source of light

in the sphere of the seen, and the source itself of truth and intelligibility in the intelligible sphere.[3]

By striving to reach the Heavens, Man was increasingly cutting himself off from Mother Earth. Thus, while in Greece Hesiod had narrated the victory of Zeus over the Titan Typhon, Gaia's youngest child, which brought about the reign of the Olympians,[4] in the Semitic Levant Yahweh defeated Leviathan, the serpent of the cosmic sea. In both cases "the lesson," said Campbell, "is equally of a self-moving power greater than the force of any earthbound serpent destiny."[5] Man was estranging himself from Earth and from the powers of fertility, perceived as "an obvious source of danger and disease."[6] The myth of Antaeus is particularly revealing: Antaeus, the half-giant son of Gaia and Poseidon, was challenging all passers-by to a wrestling contest. The odds were massively in his favour: as long as he kept physical contact with his mother the Earth, he was invincible. On his way to his 11th labour – the Garden of the Hesperides – Herakles, the solar hero *par excellence*, encountered Antaeus and quickly realized that the only way he could prevail was to lift the giant off the ground and keep him aloft. Which he duly did, choking him in a powerful bear hug. While we find here another typical achievement of a solar Superhero, the subtext could equally read as such: the disconnection from Earth kills.

The Sun, however, was less and less inclined to share power with the other Sky gods. In Egypt a first warning shot had been fired by Amenhotep IV, a.k.a. Akhenaton, who as far back as the 14th century BCE had revolutionised Egyptian cosmology by imposing the sole cult of the Sun God Aton. By occupying more and more space in the sky, the Sun was paving the way for monotheism, a move grandly facilitated by Platonism. The symbolic charge of the Sun – the *modus operandi* of solar consciousness – implied a vertical notion of power and was mirrored on Earth by the advent of the supreme ruler, of the conqueror, of the King.

But symbols are fickle and one is wise to approach them with a pinch of salt. In *Mankind and Mother Earth*, historian Arnold Toynbee related the story of Aristonicus, a Permagian prince, who revolted in 132 BCE against Roman rule by establishing a "Commonwealth of the Sun", hoping to rally slaves to its cause: "the Sun is the divine embodiment of justice. It gives light and warmth impartially to slaves and to freemen, to the poor and to the rich."[7] So the Sun, vertical symbol of kinship, could just as well be a horizontal symbol of social justice.

One may find a similar ambiguity in myths. The same story could have opposite interpretations, displaying a perplexing elasticity. In the myth of *Jonah and the Whale*, for instance: for Cashford,

> the three days of death followed by rebirth is found in many myths, from the Sumerian Inanna's descent into the Nether World, the Biblical Jonah and the Whale, to Christ's Descent into Hell. Across the ancient world the idea of resurrection was found reflected in the Moon's ever-recurring cycle.[8]

In 1889, however, American Sarah Titcombe, a student of comparative religion, had a different idea:

> The Hindoo story of Saktideva, who was swallowed by a huge fish and came out unhurt, is similar to the Hebrew account of Jonah swallowed by the whale, which is undoubtedly a sun-myth, and represents the Sun being swallowed by the earth – as it apparently sets in the west – to be cast forth by the earth again in the morning. One of the names given to the Sun was Jona, and the earth is sometimes represented in mythology as a huge fish. The three days and nights, mentioned in the account, represent the Sun at the winter solstice, when it is apparently stationary for that length of

time in the sign of Capricornus.[9]

Sun-myth, Moon-myth? A solar perspective is dumbfounded by this latent contradiction and needs to take sides. As Aristotle emphasized, A and not-A are mutually exclusive. A lunar perspective, by contrast, is quite comfortable with the paradox and has no qualms about accepting both as equally true – or false.

A similar question may be raised with Christ symbolism. Are the three days in darkness a reference to Christ's descent to Hell at Easter, as Cashford highlights, or to Christ's birth at the winter solstice, as Titcombe implies? In other words, is Christ lunar or solar in character?

In the Neolithic period a lot of attention was paid to the winter solstice, when the Sun was said to be reborn – a time, that is, when the Sun was starting his ascent higher and higher in the sky until the movement was reversed at the summer solstice. Newgrange in Ireland, 5,000 years old and counting, is testament to this reverence – the inner chamber of the monument and the passage leading to it are perfectly aligned with the Sun at the winter solstice. After a night spent in darkness, in the womb of the earth, the rays of the Sun hit and illuminate the chamber in what must be a powerful experience of rebirth – an experience that to this day it is still possible to enjoy. I, for one, would love to spend the night there.

The powerful symbolisms of the Sun and of the Sun's rebirth at the winter solstice were of course not lost on the early Christians who were, after all, coming from that very culture. In a perfectly orchestrated juggling act, the Church superseded the image of the Sun with the image of Christ as the Sun. Coming from a very Christian perspective, Hugo Rahner, a contemporary of Jung, talked of the 'dethronement of Helios', the Greek Titan and original Sun God, by which he meant the overthrow of the pagan reverence for the Sun with the image of Christ as 'the Sun of Righteousness':

> The Church is uncompromising and as she begins to take over images, words and ideas from the devotional life of the sun-worshipping Greeks, she interprets them in a manner that only has relevance to the historically clear-cut figure of her founder, Jesus of Nazareth; it is he who from the very beginnings of Christian theology is the 'Sun of Righteousness' (Malachi 4.2).[10]

One can sense the Platonic heritage in this displacement[11] – the Sun symbolic of the Good, the True and the Beautiful. For feminist Margaret Starbird, "the Jesus Christ of institutionalized Christian tradition is a male solar divinity par excellence, portrayed in the oriental patterns of the sun gods of the Mediterranean world: Ra in Egypt, Apollo in Greece, Jupiter in Rome, and Sol Invictus in Constantinople."[12] Compounding this, we may note that the day of worship for the Lord is Sun-day. For Rahner, adhering to traditional Christian exegesis, if Christ is the Sun, the Church, the bride of Christ, is the Moon. But Mary, too, is equated with the Moon:

> Mary with the twelve stars circling her head like a halo of light is like Inanna-Ishtar, who, as moon-goddess, wore as her crown the twelve constellations through which the sun – her son – moved. Like the moon, Mary became the Mistress of the Waters, guardian of the rhythmic ebb and flow of the womb.[13]

Case closed then? Hardly. For indeed the motif of death and rebirth, sometimes through a 3-day descent in the underworld (but not always), is essentially lunar in character and percolated throughout Antiquity. "All dying and resurrected Gods were once gods of the Moon,"[14] said Cashford. This was expressed in the figure of the twice-born god – undergoing death and dismemberment followed by a rebirth – as illustrated by the myths of Dionysus, Attis and Osiris. And by Christ too, complete with the battle with the sea monster:

> Since the Moon could disappear for two or three nights, Jesus' nights of death become two nights, rising on the third day. As the lunar gods had illuminated the dead in the underworld, sometimes slaying the monster of darkness, so Jesus is often shown releasing the souls from death by spearing the monster of hell, imagined as a whale... Jesus is here cast in the role of the Bright Moon slaying the dragon of the Dark Moon, bringing rebirth.[15]

So we may be left with the same puzzling conundrum as with the myth of Jonah. Sun or Moon? Yet, in genuine soli-lunar fashion we may also see a different reality emerging, one that blends both viewpoints. This shift of perspective was clearly articulated by Alan Watts:

> In the cycle of the Christian Year the rites of the Incarnation are governed by the solar calendar, since they are connected with the Birth of the Sun, and so fall upon fixed dates. On the other hand, the rites of the Atonement, of Christ's Death, Resurrection and Ascension, are governed by the lunar calendar, for there is a figure of Death and Resurrection in the waning and waxing of the Moon.[16]

Or, as Cashford puts it beautifully: "In the language of metaphor, Christ is born as the Sun, but is reborn as the Moon."[17]

At the risk of being suspected of displaying the disturbing signs of a Christ complex, I do feel a personal resonance with that statement. Referring back to my astrological chart and my Sun/Moon see-sawing, I feel that I was born on the side of the Sun.

My childhood was spent with three brothers and my pursuits were very solar in character. I studied science then business. I ended up in a job – oil trading – where 90% of the workforce was masculine, a job where achievement, results and performance were paramount. A job of numbers, of reason, detachment and clear analytical thinking. Apollo ruled. A job at the heart of our economic model – feeding the Beast with the blood of the Earth. I am schematizing of course. It was also a job where intuition played a big part.

And then in my 43rd year I was dismembered. Profoundly shocked and stirred from my slumber. In truth, however dramatic, it was the perfect experience. I had wanted to leave my job for a while but was not able to gather the courage to do so. In many respects it was a very enviable job and in no way could I give myself permission to leave it. Millions of people would love to be in my shoes, I felt; would not resigning be a damning display of ungratefulness, an unforgivable act of hubris? And to do what anyway? I was 42, had three children to raise and my only professional competence was oil trading – and if I was ever to leave, it was certainly not to do the same job elsewhere. A Catch-22 situation. I am convinced that I somehow *created* my short experience in custody so as to liberate me. But to which 'I' am I referring here? Certainly not the frightened, timid 'I' of day-to-day life. This 'I' would have stayed forever stuck in fruitless mind games. No, what I am talking about is a higher 'I', also called the Higher Self, or the *daemon*. A tutelary authority that watches over you and is not afraid to take the difficult decision in your place. I found freedom in my cell. There I found the courage to leave and start afresh, whatever difficulties might lay ahead. I suddenly realized on my third day in the underworld that I had been given a sign, a sure sign that it was time to extirpate myself from my job and be born to a new one – although what form this new life was to take, I had not the slightest inkling. I found a firm resolution then, that was not to leave me for even a split second afterwards.

I remember a particular episode of that experience, an episode that shows that even in our darkest hour, when everything seems blurred and frightening, the universe – for want of a better word – is still speaking. It was the end of the second day. I had been interrogated until midnight the previous day, had spent the night in a small cell, and had been interrogated again the whole day. It must have been 7pm and I was going to be transferred to another jail – the Conciergerie on the 'Ile de la Cité', a notorious historic jail in the middle of Paris, much used during the Revolution. There I was going to spend the night before being presented to the investigating magistrate the following day. I was very frazzled. The police officers who had interrogated me were, however, quite sympathetic to my plight. As I was about to leave and be taken away, one of them, a woman, gave me a small chocolate bar to comfort me. I was very touched by this gesture – in these circumstances, any crumb of kindness is a feast – but today I cannot but not notice the symbol that was screaming in my face, although I did not see it at the time: she handed me the famous coconut chocolate bar going by the name of 'Bounty'. A bounty indeed these three days were. Which does not mean that they were easy, nor that the experience only lasted three days. It lasted in fact seven years, which was the length of time required for the trial to eventually go through – it was a long procedure and many people were involved – until I was cleared of any wrongdoing. That's a long time with a Damocles sword above one's head. It had been a long journey, the turning point of which had been my encounter with the Dark Goddess, when I had seriously considered how best to end my days if things were really going pear-shaped. I had in mind the words of Hunter S. Thompson, who had his share of encounters with the Law. Being involved with the justice system was like riding a wild stallion, he remarked: things had a habit of taking a life of their own and there was very little that one could do about it.

When I left the judge on that Thursday evening, I smelled the
perfume of the street as never before. Then I treated myself to
a nice meal. I felt all grey and greasy from three days of being
locked up but something had been awakened in me, a mix of fear
and anticipation. I was reborn and, while it was by no means
apparent to me then, I was reborn as the Moon.[18]

In astrological symbolism, 42 years is the age of the Uranus
'half return'. Uranus orbits the Sun in 84 years so after 42 years
it occupies a position opposite to its position at the time of
birth – a geometrical half-circle, referred to as the half return.
Uranus is the planet of unexpected surprises. It was the first
'outer' planet to be discovered. Astrologers pay a lot of attention
to the synchronicity of the discovery of a new celestial body
with outer events in the human sphere. The year was 1781, a
period of intense social changes, in between the American and
the French revolutions. After millennia of a sky filled with seven
planets, suddenly a newcomer entered the scene. It was indeed
a revolutionary discovery, shattering the old Ptolemaic model.
The Universe immediately expanded massively. A new God
had arrived, a God of rupture, of sudden change, of upheavals,
an iconoclast who was not afraid to destroy the status quo
with one swoop of the arm. Uranus strikes suddenly, without
warning, often with painful results – not for the sake of inflicting
gratuitous pain, but to *awaken*. Its symbolism is similar to the
Tarot card of the Tower – Arcana 16 – when the characters are
suddenly expelled by a lightning strike from their tower. The
half return is a period when Uranus is felt intensely. This is
the age of the mid-life crisis. Compounding this, Uranus is in
my natal chart in opposition (180 degrees) to my Sun. At the
half return, therefore, Uranus was conjunct with my Sun – and
astrologers all over the world are well aware that the transit
of Uranus over the Sun is also potentially a very challenging
period. A double dose of Uranus medicine for me, then, at the
age of 42. Clearly, I got expelled from the tower where I was

resting, encased as I was in the brick wall of my lopsided mind constructions. Uranus' lesson was painful but necessary. It might be added that if Uranus is opposing my Sun in the natal chart, it is also conjoining my Moon, which stands at the opposite end of the Sun – Uranus was therefore aligned on my Full Moon, one of the main protagonists of this book. In that sense it is no real surprise that this planet should have been the agent of change for me: Uranus jolted my Full Moon like lightning. Born as the Sun I was reborn as the Moon, Uranus switching the polarity in its customary harsh but truthful ways. In the astrological chart of the Moon landing, Uranus is also conjoining the Moon, as it is in my natal chart. And what better symbolism to express the sheer jolt experienced by the whole world when these two men were suddenly seen walking on the Moon – what an awakening for mankind and, arguably, for the Moon herself!

Back to our historical thread, though. The world in the four centuries following the death of Christ was a brewing cauldron. Many sects, creeds, philosophies and spiritualities were vying for supremacy. One of the main competitors to the nascent Christian faith was then Mithraism, a religion that was extremely popular with the Roman legions and that was articulated around a powerful solar imagery. "Nearly all of the Sun-gods," said Titcombe, "are represented as having been born in a cave or a dungeon."[19] This was certainly the case for Mithras, who was born from a rock in a cave: "the most common representation of [Mithras'] birth," said German scholar Manfred Clauss,

> shows Mithras naked, his sole clothing the phrygian cap; and wielding a torch and a dagger. His most important exploits are thus adumbrated already at the birth. With his torch he brings light: he is *genitor luminis*, creator of light, and, as sun-god, himself also that light. With the aid of the dagger he creates light, by killing the bull.[20]

The killing of the bull by a solar god or by a solar animal is a major archetypal motif. With its horns symbolising the crescent Moon, the bull was indeed a lunar animal, dear to the Goddess. For Eliade, "the bull's horns were long ago compared to a crescent and likened to the moon... idols in the form of an ox, often connected with the cult of the Great Mother (the moon) are frequent in the Neolithic Age."[21]

Campbell contrasted the bull with the lion, and the serpent with the eagle. In this double polarity he saw the expression of Sun/Moon symbolism:

> As the bull is symbolic of the moon, so the lion, with his great radiant solar face, is the symbolic animal of the sun. As the rising sun quenches the moon and the stars, so the lion's roar scatters the grazing animals, just as the lion's pouncing on the bull symbolizes the sun's extinguishing the moon. If we recall the serpent, we recognize the eagle, the solar bird, as its counterpoint. So we have these parallels: eagle against serpent, lion against bull, sun against moon.[22]

Carl Kerényi noted that bull, serpent and the Feminine were chief features of the lunar god Dionysus: "To the Greeks, Dionysos was pre-eminently a wine god, a bull god, and a god of women. A fourth element, the snake, was born by the bacchantes, as it was by... goddesses and priestesses of the Minoan culture."[23]

Mithras' ritual killing of the bull can therefore be seen as the enactment of the Sun's victory over the Moon. Is the corrida a direct descendant of this sacrificial ritual? The torero in his shiny golden suit, the embodiment of the Sun, facing the symbolic animal of the Moon – a death ritual and, for Federico Garcia Lorca, a 'bellow of pain':

> That bellow of pain comes from the frenzy of the bullring, and it expresses an ancient communion, a dark offering to the

Tartesian Venus of the Dew, who was alive before Rome and Jerusalem had raised their city walls. It is offered in sacrifice to the sweet mother-goddess of all cows, queen of the Andalusian bull ranches, all but forgotten by the civilization which now stands near the lonely salt marshes of Huelva.[24]

The ability of the snake to shed its skin and grow a new one made it an obvious candidate for Moon symbolism. From very early on, the serpent was sacred to the Goddess(es), a symbol of fertility and renewal. It famously became demonized at a later stage, as evidenced in the Biblical story of the Fall. Its crawling ways similarly brought an obvious association with the Earth. For the Aztecs, the surface of the Earth was imagined as an immense interweaving of reptiles. The Earth was Serpent Woman and she was referred to as Coatlicue, she of the Skirt of Serpents. The modern resurgence of research into Earth energies, ley lines and power lines criss-crossing the landscape has revived serpent symbolism. One of the most well-known books on the subject is *The Sun and the Serpent* by Hamish Miller and Paul Broadhurst, an investigation into lines of energy running through the land.

The archetypal fight between the eagle and the serpent was a staple of pre-Colombian civilizations. In 1325, after decades of wandering, the Mexicas – more commonly known as the Aztecs – reached a lake in an upper valley of Central Mexico. On an island close to the shore they saw an eagle, perched on a cactus, holding a snake in its beak. This was immediately interpreted as an omen sent by the gods and on this island they founded their main city, Tenochtitlan. The Mexicas not only gave their name to the country, but its mythical foundation too: on the Mexican flag an eagle is conspicuously represented with a snake in its beak. The two main temples in Tenochtitlan were built in honour of Tetloc, the god of rain and storms, and Huitzilopochtli, the Sun god of war. Long ago, Huitzilopochtli had to fight a mythical battle with his sister Coyolxauhqui, the Moon goddess, who was plotting to prevent his birth. At the bottom of the temple a magnificent giant stele of stone was placed by the Aztecs, representing Coyolxauhqui, defeated and dismembered. Two centuries later, Hernán Cortés put an end to the Aztec empire in the name of the Spanish Crown. Cuauhtemoc, the last ruler, was tortured to reveal the whereabouts of the Aztec treasure. The main temples of Tenochtitlan were duly pillaged, their stones used by the new rulers. A cathedral was erected on the emplacement of the pyramid of the Sun. The vestiges of Tenochtitlan can be seen to this day in the centre of what has become Mexico City. It is a poignant sight: the remnants of what must have been a staggering temple complex are scattered about, skeletons of a giant being whose bare bones of stone lie exposed under the relentless sun and in the honking of cars.

In the Museum of Anthropology of Mexico City, one reads that "the Prehispanic people considered the light and warmth of the Sun to be equal to life itself. Therefore their creation myths saw in the presence and destruction of that star the precarious nature of life and the need for men to help maintain the Sun as the supreme deity." In Aztec cosmogony, four Suns were created in turn, each one associated with an element – first earth, then wind, fire and finally water. Five was the quintessence, the centre, the realization of the first four principles. The fifth Sun was created "by two numena, Nanahuatzin and Tecuciztecatl, who became the Sun and the Moon respectively. For this act of generation to take place, both gods had to set themselves on fire." This act of self-sacrifice meant that man had a terrible debt towards the gods – a debt that had to be repaid in blood, one's own and one's enemies. The magnificent Stone of the Sun, also visible in the Museum, records the birth of the fifth Sun. It is a powerful testament to the absolute awe with which Mesoamerican civilizations regarded the Sun.

Deep into the jungles of Chiapas and of Guatemala, or on the lowlands of Yucatan, Mayan cosmology, too, was deeply informed by Sun and Moon. According to Mexican anthropologist Oswaldo Chinchilla Mazariegos, former curator of the *Popol Vuh* museum,

> mythical narratives devote considerable attention to explaining the reasons why one of the heroes eventually became the sun, and the other, the moon, and the related questions of why the light of the moon is weaker... The distinction is clear in stories that describe the sun and the moon heroes, respectively, as male and female, although the gender contrast is not always present, and there are numerous myths in which both the sun and the moon are male.[25]

In one myth, the moon was as bright as the sun. People could not sleep at night – until the sun dimmed her light by plucking out one of her eyes. In the Chiapas Maya myths, the moon was the sun's mother – a noticeable parallel to Mary, Mother of Christ. The solar heroes were usually young males who had to overcome family hostility in the form of their half-brothers and, especially, of their grandmother – a reminiscence of Marduk's fight with his female ancestor – who was often portrayed as possessing an evil character and had to be vanquished, usually through sexual assault:

> These old goddesses overlapped considerably with... primeval monsters... Best known among the latter are the great birds... often coupled with feminine, terrestrial counterparts who were no less threatening, and were finally overcome by the heroes. Their defeat was crucial for the rise of the luminaries and the coming of humanity.[26]

The Mayas were admirable astronomers and builders, as numerous sites in Mexico and Central America attest. Their cosmology revolved around the *Hunab Ku*, the supreme creator whose representation is strongly reminiscent of the yin/yang symbol: a black and white motif incorporating the square and the circle, symbolising the dance of polarities whose reconciliation in the centre generates harmony. In Chichen Itza, the great Mayan city located in Yucatan, stands the Pyramid of Kukulcan, a 26-metre high structure. Kukulcan was the Mayan version of Quetzalcoatl, a creator god in the shape of a feathered serpent, whose worship was said to have been introduced to the Mayas in the 5th century by representatives of the extraordinary city of Teotihuacan, near Mexico City. With its hybrid form, half-bird, half-snake, the feathered serpent may be said to reconcile Heaven and Earth. The Pyramid of Kukulcan was made of a double structure: underneath the external pyramid stood an inner, smaller, one. The two pyramids were built to seal the knowledge of the Mayas: the external pyramid was said to be the pyramid of the Sun, guardian of outer and exoteric knowledge, while the inner pyramid was the pyramid of the Moon, guardian of the secret, of the occult, of esoteric wisdom.[27] The numerological symbolism of the outer structure pointed to the *Haab*, the 365-day yearly, secular calendar, while the inner structure was built around the symbolism of the *Tzolkin*, the 260-day sacred calendar. The Mayas were obsessed with equilibrium and balance, and nowhere is this more apparent than in the exquisite architecture of the Pyramid of Kukulcan, with its Sun and Moon symbolic union.

Further south, the many cultures that flourished in the Andes were no less mythically informed by the Sun and the Moon, with the Sun occupying a central role. West of La Paz, in Bolivia, lie the remnants of the great city of Tiwanaku. The Tiwanaku culture encompassed a vast expanse of time, from 1000 BCE to 1200 CE, when it was eventually absorbed into the Inca empire. In the local Aymara language, Tiwanaku stood for "Sons of the Sun". A massive temple was built in homage to the Sun, as well as a striking Puerta del Sol – a door of the Sun – probably used for ceremonies and rituals.

In May 2019, as I was putting the last brushes to this book, I went to Titicaca to visit the mystical Islands of the Sun and of the Moon. This was to be a closing of the loop for me and I felt it was a fitting end. On my way to the lake, as I was passing through La Paz, I heard about the Valle de la Luna, the Valley of the Moon, located a mere ten kilometres outside the city. My curiosity was tickled. Surely the name referred to some mysterious connection to the queen of the night? I proceeded to go there with a guide and asked him about the origin of the name, expecting an esoteric and highly symbolically charged explanation. It was not to be: while Neil Armstrong was in space, the guide proceeded to explain, he noticed a very large white formation, the Salar de Uyuni, in the south of Bolivia, and he was intrigued. Two years after the Moon landing, Armstrong decided to check for himself. While in La Paz before travelling down to Uyuni, he was taken to the Valle de la Luna. At that time, though, the Valley was known as a Valle de las Animas – a valley of the souls – in reference to the souls of dead local people said to reside in the spiky stalagmites typical of the site. The landscape of the Valley, remarked Armstrong, was very similar to the Moon surface, and lo and behold! the valley henceforth became known as the Valle de la Luna. I was dumbfounded: I was expecting Moon symbolism and I was encountering a Moon veteran. Symbols are elusive and tracking them can yield ironic results. I was certainly not expecting to walk in the footsteps of Armstrong outside La Paz, yet, in the context of this book, I took it as a nod that I was on the right track. A little pebble on the road.

Echoing the Mexicans thousands of kilometres away, the iconic Sun-worshipper Incas also mythologized the fight between eagle and serpent. "The Andes were bedazzled by the Sun,"[28] said historian William Sullivan and for the Incas, "the mythical significance of the raptorial birds is their ability to soar or dive – like the sun – to the limits of the world of the living, that is, to the gates of heaven itself."[29] The counterpoint to the eagle, the serpent, "was symbolic of tectonic upheaval on earth, and 'uranic' upheaval in the domain of the 'celestial earth'."[30] In *The Secret of the Incas* Sullivan relates an Inca legend, in which "Pachakuti Inca, in desperate battle with a fierce tribe of the eastern forest, saw his army set upon by a gigantic serpent. Pachakuti raised his eyes to heaven in supplication, and immediately a great eagle swooped down, picked up the serpent, and smashed it to death on the rocks below."[31]

It is certainly no bed of roses to have been born a serpent in the last three thousand years of Western history. I, too, have had my own brushes with snakes and these encounters go from the ridiculous to the sublime. My first encounter happened at 14 at my grandmother's home in the countryside. Grass snakes were often seen in the garden and my grandmother was terrified of them. On that particular day, one hot summer afternoon, she spotted a grass snake lazily coiled in a garden armchair. It was a big one, of a brown/greenish tinge. "Philippe, you have to kill it!" I was alone with her in the garden and I suddenly felt invested of a mission. At 14 you are no man but you want to be one, or be seen as one – without of course having the slightest clue of what that means. Decades later, you are often none the wiser. Anyway, what followed was very archetypal and very sad. I felt full of male courage, in combat mode, ready to fight the beast that was threatening my grandmother, donning the mantle of the Saviour, rescuing the Feminine. I was Apollo fighting the Python, the carrier of murderous values, man vs. nature, a one-sided mentality of aggression in the name of protection. I killed the snake with stones, an agonizing process that sends me into whirling sadness whenever I think of it. "Forgive them, for they know not what they do," Christ had said. May *you* forgive me, my brother, I was an unconscious fool.

Three years ago, I participated in a week-long 'process' in South Africa organised by the Gaia Foundation. The process was led by two white Sangoma witch doctors. We were staying in a wilderness retreat three or four hours from Cape Town, in the Langeberg Mountains, a region known for its very unique ecosystem. On the second day we were asked to leave the retreat and head out, alone, anywhere we felt called to go. The script was just to rest in nature, without doing anything in particular, or without expecting anything – just opening to what may come and, above all, accepting to feel vulnerable. A river was flowing twenty minutes or so away from the camp and I decided to venture that way. As I was walking I noticed a stone on the path, and it looked strikingly like a serpent's head. I instinctively felt that this was a sign that I might encounter a snake in my wanderings, although the voice of Reason, the voice that refuses to believe, immediately dismissed it as nonsense. I kept going and reached the river, which was bigger than I had expected. In the middle laid big flat rocks, encasing pools of clear water, and I felt called to go there. While stepping over stones to reach the pools, I sensed something moving under me. I turned around and there it was – a magnificent water snake, half-asleep. I had obviously stirred it from its afternoon slumber. I moved away rather quickly. The thing is, when you are in an environment that you do not know, how can you tell which animal is potentially dangerous or not? Erring on the side of caution I stayed at a respectable distance, not too close but not too far either so I could have a good look at it. After a short while the snake decided it had enough and it moved away, gliding gracefully downstream in a typical sinuous way until I lost sight of it. I reached the rock in the middle of the river, stripped down and, stark naked, bathed in the pools of cool water – keeping all the while a watchful eye over the river upstream, attentive to further snaky appearances. I stayed a couple of hours on the rocks, taking regular dips, leaving sun and wind to dry my skin, an Adam of a man in the Garden of Eden, vulnerable yet wonderfully alive. The Tree of Life was all around me, in glistening, colourful shapes and sounds. I had never done that before – be naked in nature, a proper animal relying on its senses, relaxed yet alert. Amazing really what it does to you, this nakedness in the open: this openness in the nude! Time was up, however, and I started to walk back. As I was about to reach the retreat, I came across a bend in the path and was suddenly jolted from my reverie by a startling sight, right after the bend: an orangey cobra, holding in its mouth a dead black-and-white snake. It looked like I had again inadvertently interrupted Mother Nature in her daily rhythm. The cobra, sensing me, immediately raised its body, dropping the dead snake to the floor, and after a few seconds of stillness, quickly disappeared in the bushes. I was mesmerized. In the morning, the witch doctors had introduced us to the Southern African concept of *Sereti*. To have *Sereti* is to have presence, power, authority, to exude a sense of threat. It comes from a deep inner place, from a profound alignment of oneself with oneself. When the cobra raised its head, I immediately knew what they meant. This cobra had *Sereti*, lots of it. You do not mess with *Sereti*; startled, you take notice and acknowledge. Three snakes in the afternoon, then, two alive and one dead. What to make of it? When we shared our stories later on, of the 18 participants no one had encountered any snake. I, alone, had met not only one but three, and everybody sensed that it was very significant. In which way though? 'Three' is certainly an important number. In astrology, three signatures of a theme in a chart flag up a key aspect of the chart. In shamanic journeying, meeting the same animal three times is highly significant, too. To meet the serpent three times in waking life is just as meaningful and, in keeping with the traditional attributes of the snake – death and rebirth, renewal, shedding one's skin – carries a sense of a major turning point. Which was indeed the case, having poignantly ended a 32-year relationship six months before. "All teems with symbol," Plotinus had said.

At first sight, lion symbolism seems rather straightforward, as academic Angela Voss highlights through the esoteric doctrine of correspondence: "The ray which crystallises as 'Apollo' on the level of Ideas, will bring forth the 'Sun-soul' on the cosmic level, the physical Sun on the material, the King on the human, lion on the animal, heliotrope on the vegetative and gold on the mineral."[32] Echoing Campbell, Liz Greene sees a polarity between lion and serpent, "a *unio oppositorum*, in the iconography of the Mithraic lion-headed god Aion."[33] Yet the lion was once associated with the goddess, says esotericist Arthur Versluis: "the ancient goddess of the mountains who had many names (Rhea, Kybele, Agdistis) was often accompanied by or incorporated the symbolism of the lion."[34] Artemis presents an interesting case, being associated both with lions and with serpents. Maybe this is in keeping with what scholar Sorita d'Este, in her study of the Goddess, describes as her dual nature – lunar but also, maybe less obviously, solar: "the twin serpents motif is... seen in association with Artemis, on coins and at temples in friezes and statues,"[35] while at Ephesus, "one of her primary temples, we find depictions of lions."[36] She is also described by Pausanias as holding a lion in her left hand and a leopard in her right, which, according to Sorita d'Este, is suggestive of her association with the gods most important to her – Apollo with the lion and Dionysus with the leopard, again showing an affinity with solar and lunar principles, both male gendered.

The Spanish conquistadores who ruthlessly made their way into America were in all likelihood far less sensible to the symbolism of Sun and Moon, yet their appetite for gold and silver – respectively, the Sun's and Moon's metals – made them no less subject to the mythical realm of the luminaries. In their search for the elusive Eldorado, gold would of course occupy their chief attention. During the same period, another group of adventurers was eagerly searching for Gold. For the alchemists, the goal of the Opus was the production of the fabled philosophical stone, leading to the alchemical gold. This was achieved through the *hieros gamos*, the union of the masculine and the feminine, of *Sol y Luna* – of Sun and Moon. But despite this balance it was gold that alchemists were after, not silver, and gold was without question equated with the Sun, as 16th century alchemist Michael Maier, quoted by Jung, clearly highlighted:

> The sun is the image of God, the heart is the sun's image in man, just as gold is the sun's image in the earth… and God is known in the gold. This golden image of God is the *anima aurea*, which, when breathed into common quicksilver, changes it into gold.[37]

Was the primacy given to the Sun a reflection of the fact that alchemists were in their vast majority male?

By the time of the alchemical quest, the respective genders of Sun and Moon were indeed quite clearly masculine and feminine, with a few exceptions, and had been so for many centuries, if not millennia. This widespread association Sun/Masculine and Moon/Feminine is today totally abhorrent to many feminists, who see in it a patriarchal plot to keep women in their (lower) place since in our culture, of course, the Sun has primacy over the Moon. One such author is Janet McCrickard, who explores in *Eclipse of the Sun* the rejection of Sun Goddesses by a male mentality that "reject the Sun-goddess for the same kinds of reason that they still want to exclude women from being priests, judges or members of the Kennel Club. It's just not a proper job for the girls."[38] In the initial stages of my reading I felt McCrickard had a point and I engaged with the book with a fair amount of sympathy, but little by little something seemed amiss. Instead of challenging the dominance of the Sun she was endorsing it and pushing women to become more solar. But by doing so the risk is to create good "daughters of the patriarchy", and the conclusion of her book seemed to confirm exactly that. She finished with a flourish against intuition, feeling and emotion as "cowardly bypaths of self-delusion."[39] Only science, she concludes, can show the true way to "improve and redraw the maps of reality"[40] and with a final movement of the chin she asserts, no doubt under the applause of many men, that

> the Earth and the human race itself will not be saved by anathematizing scientific inquiry from the safety of some One True Religion position, nor by moonbathing, Tarot card-reading, or pretending to be Neolithic woman, but by *breadth of understanding*, by rationally communicating with each other, and by rational solutions.[41]

My heart sinks when I read such statements, whether they are written by a man or by a woman. McCrickard is stuck in a typical 'either/or' solar attitude that cannot accommodate 'both/and'. McGilchrist, whom we will encounter later, would see in it a typical left-brain *oukaz*. The denial of the Imagination and of the symbolic attitude, the evangelizing faith in the scientific inquiry, called scientism, the total abdication to Reason as the only valid epistemology – this is the denial of the Moon. Not the Moon as in 'woman', but the Moon as in 'lunar consciousness', a consciousness that is accessible to both men and women, just as solar consciousness is accessible to women and men. To put it symbolically it is not more Sun that we need, as McCrickard rather forcefully advocates, but more Moon – for men and women – while we must still of course recognise both the solar feminine and the solar masculine. We need to redress an imbalance, not run away with the Sun, who has run far enough.

But she certainly makes a very valid point: the Sun/Masculine-Moon/Feminine couple runs deep in Western culture. The founding text of the esoteric arts and of Western hermeticism, the *Tabula Smaragdina* – the *Emerald Tablet* – dating from the very early part of the Common Era and supposedly written by the mythical Hermes Trismegistus, was unequivocal: "The father thereof is the Sun, the mother the Moon. The Wind carried it in its womb, the Earth is the nurse thereof. It is the father of all the works of wonder throughout the whole world." In the 13th century, St Francis addressed a canticle to 'Brother Sun, Sister Moon'. In the *City of the Sun*, a utopian tale written by Dominican philosopher Tommaso Campanella in 1602, "the Sun is the father, and the earth the mother"[42] and, as the father, the Sun reigns supreme: "Beyond all things, they venerate the sun, but they consider no created thing worthy the admiration of worship. This they give to God alone... They contemplate and know God under the image of the Sun, and they call it the sign of God."[43]

The esoteric tradition of the West owes a lot to the *Tabula Smaragdina*, as it does also to Jewish mysticism and to the Kabbalistic Tree of Life. And I find it hard not to see in esotericism a strong bias in favour of the Sun, a perfect reflection of our cultural prejudice. In the Theorem IV of the enigmatic treatise *Monas Hieroglyphica*, written in 1564, legendary occultist John Dee, royal astrologer to Queen Elizabeth I, could not put it more clearly:

Although the semi-circle of the Moon is placed above the circle of the Sun and would appear to be superior, nevertheless we know that the Sun is ruler and King. We see that the Moon in her shape and her proximity rivals the Sun with her grandeur, which is apparent to ordinary men, yet the face, or a semi-sphere of the Moon, always reflects the light of the Sun. It desires so much to be impregnated with solar rays and to be transformed into Sun that at times it disappears completely from the skies and some days after reappears, and we have represented her by the figure of the Horns (Cornucopia).[44]

In astrology, the Sun is masculine and represents the father, the Moon, feminine, standing for the mother. While both luminaries occupy a central position in astrological lore, it is the Sun that stands out, being at the core of the personality, the essence of one's being, around which all the planets revolve. The Moon reflects the light of the Sun, a recurring theme in Western occultism and a fact that more than any other has contributed to her demotion. In *The Lunation Cycle*, Dane Rudhyar, the father of humanistic astrology and a very delicate and sensitive writer, explores the interplay of Sun and Moon and in so doing sticks to the traditional canon:

> The light of the Sun is dazzling in its brilliancy, and its penetrating warmth imparts to the organisms which strive to rise from the gravitational level of the earth-surface the power necessary for their growth. The light of the Moon is distant and cool. It resembles the image of a lamp in a mirror – a glow which tells us accurately enough what things are, yet which does not go forth to bring the vital essence and warmth of these things to us, as does the light of the lamp itself.[45]

Rudhyar has written exquisite pages on the lunation cycle and its psychological symbolism. Crucially, though, he is keen to emphasize that the Sun and Moon dance, however compelling it is, is meant to take us back to our very own Earth:

> The basic factor in the lunation cycle is neither the fiery, effulgent and seed-releasing sun, nor the cool and objective, concept-building or body-developing moon; it actually is the earth whose need demands the cyclic interplay of the solar and lunar activities.[46]

In the Tarot deck, a direct descendent of the 22 paths of the Tree of Life, the Moon – Arcana 18 – carries a rather heavy connotation, as described by esotericist AE Waite:

It illuminates our animal nature, types of which are represented below – the dog, the wolf and that which comes up out of the deeps, the nameless and hideous tendency which is lower than the savage beast. It strives to attain manifestation, symbolized by crawling from the abyss of water to the land, *but as a rule it sinks back whence it came* (my italics).[47]

By contrast, Waite highlights the spiritual redemption implicit in the Sun card, Arcana 19:

It is the destiny of the Supernatural East and the great and holy light which goes before the endless procession of humanity, coming out from the walled garden of the sensitive life and passing on the journey home.[48]

The Arcana numbers of Moon and Sun – 18 and 19 – are quite puzzling. Number 19 is reminiscent of the Metonic cycle, previously encountered. Besides, the Moon's nodes[49] have a cycle of 18.6 years, straight in the middle of the two numbers. And the Saros cycle – the cycle of the eclipses of the Moon and Sun – is 18 years. Were numbers 18 and 19 in the Tarot deck assigned to Moon and Sun consciously then, to reflect their dance? Or, even more mysteriously, unconsciously?

Waite, the co-creator of the Rider-Waite Tarot deck, the first to illustrate all cards, joined for a couple of years the main order of the esoteric revival of the late 19th century, the Order of the Golden Dawn – a name that is doubly solar.

The exaltation of the Sun may never have been better articulated than by Marsilio Ficino in the Renaissance. Ficino, a towering figure who reintroduced Plato to Western audiences, directly borrowed in his *Book of the Sun* the Sun metaphor from the Greek philosopher and placed it in a Christian context: "the light of the Sun is similar to Goodness itself, namely, God."[50] Much as Rudhyar would state five hundred years later or so, and as the Neoplatonists did 1,200 years before, the role of the Moon was that of distributing the light of the Sun back to Earth: "the Moon, the lady of generation, has no manifest light except from the Sun. When she is in perfect harmony with the Sun, she takes from it all the celestial powers which are gathered there, as Proclus says, so that she may convey similar powers down to our earth."[51] Ficino, a true magus who was working magically with the symbols, may arguably be credited with being the first one to use astrology in a psychological way, away from the 'petty ogres' of his time whom he despised for their deterministic use of astrology. As Voss remarks in her study of Ficino:

> In the esoteric traditions the first step towards... gnosis is the apprehension of the hidden correspondences in the material world... As a symbol-system, [astrology] clearly is not necessarily restricted to the observation of 'effects' on human beings... If planets are also gods, they can be appealed to, and negotiated with, as in divinatory ritual. A horoscope then becomes a dynamic play of divine energy, not a static and immutable blueprint.[52]

Ficino saw himself afflicted, both psychologically and physically, by the planet Saturn, which was placed on his Ascendant – that is to say, Saturn was rising on the horizon at the time of his birth. As such he would doggedly engage with it so as to turn this affliction into a boon. Saturn became for him the planet of philosophy but he could never shake off a somewhat melancholic mood. As a Christian, Ficino was, however, treading on thin ice. Ever since Augustine, Christianity and astrology had had a fractured relationship and even though for Ficino, "the mysterious realization of another order of reality could occur with equal intensity in the ritual of astral magic or the Christian mass,"[53] he had to be cautious with his choice of words. The shadow of the Church was lurking in the background. In the conclusion of *The Book of the Sun*, Ficino leaves no doubt as to where his loyalties rest:

> According to Plato, [Socrates] called the Sun not God himself but the son of God. And I say not the first son of God, but a second, and moreover visible son. For the first son of God is not this visible Sun, but another far superior intellect, namely the first one which only the intellect can contemplate. Therefore Socrates, having been awakened by the celestial Sun, surmised a supercelestial Sun, and he contemplated attentively its majesty, and inspired, would admire the incomprehensible bounty of the Father.[54]

Chapter 3

Apollo

The Evolving Image of the Sun-God

May I not forget
Apollo the Archer
The gods tremble at him
When he enters the house of Zeus
– Homeric Hymn to Apollo[1]

Ten years ago I became totally engrossed in a TV series called *Battlestar Galactica*. The synopsis goes like this: in an unknown time, humanity, spread out on different planets named after the twelve signs of the zodiac (the Twelve Colonies), has developed a very fractured relationship with robots and artificial intelligence (an increasingly classic theme). The robots are actually more than robots: they are a cybernetic sophisticated lot called Cylon – an acronym for Cybernetic Lifeform Node.[2] One fine morning, when everyone is busy sipping their flat whites, the Cylons launch a devastating nuclear attack and destroy the whole human race… or almost all of it. A few survivors, who happened to be scattered on various spaceships at the time of the strikes, miraculously escape Armageddon and gather together around the last remaining battleship, the ageing 'Battlestar Galactica'. Desperately trying to escape the clutches of the Cylons, they are a tribe of 50,000 wandering Jews placed under the unlikely leadership of Laura Roslin, an ex-teacher who finds herself suddenly thrown into the role of President of what's-left-of-humankind, and of Commander Adama, in charge of the 'Battlestar Galactica'. In Hebrew, *adamah* means earth. 'Roslin' is reminiscent of the mysterious Rosslyn Chapel south of

Edinburgh, with its Templar connections. It also brings to mind the 'Rose Line', the wise Feminine. Guided by vague Scriptures, they place all their hopes in the search for a mythical planet, a planet that nobody knows whether it exists for sure, a planet where they can start afresh – the planet of beginnings, the 13th colony, where they are said to originate from: our very Earth. *Battlestar Galactica* is a story of homecoming. Even the Cylons want to come home, in their desperate search for their Creator and for their soul. I love that series, I call it the Mother of All Series. I find in it all my pet subjects.

On board the 'Battlestar Galactica', jet Viper pilots are busy defending a human race about to be consigned to the dustbins of history. The best of them all is Adama's son, Lee, who goes by the handle name of Apollo. Lee has a dark side, having lost his mother in the Cylon attack, and his brother, killed as he was training to be a pilot a few years before. He is also in conflict with his father, whom he holds responsible for his brother's death. But Lee also displays all the qualities that are associated with Apollo: excellence, clarity of mind, charisma, radiance, reliability, rationality, courage. A solar character, a beacon.

When the Greek God Apollo was entering Olympus, all the Gods and Goddesses were rising up from their seats to greet him. Such was his clout. One might say that Apollo had *Sereti* in abundance. His mother Leto was the only one to remain seated, relieving him of his golden bow and arrows as he was taking his place among the Olympians. Leto and her children, Artemis and Apollo, were a very tight unit, a bond that found its origin in the story of their births. Leto was the daughter of the Titans Coeus and Phoebe ('the Moon'), and therefore a granddaughter of Gaia, the Earth. She became pregnant by Zeus after her beauty caught his eye but in the process, as so many other female deities or mortals, she attracted Hera's wrath. Hera, Zeus' spouse, swore that Leto would never find a place to deliver her babies. In her wanderings Leto was said to have visited all the mountains and islands of Greece until she eventually came to the little island of Delos, a rocky and unfruitful crag in the Aegean sea. Despite being promised riches by Leto, Delos was reluctant to give sanctuary to the Goddess, for it had been prophesised that Leto's son would be of a terrifying power. Leto swore by the Styx that Apollo would build his first temple on the island, which would make Delos a sacred place – an oath that the God dutifully honoured. Apollo's birth was a long, protracted affair that lasted nine days. In her jealousy, Hera was keeping Eileithyia, the divine midwife, hidden away in Olympus behind golden clouds so that she was unable to assist the delivery. Eventually Eileithyia was informed by Iris about the events taking place at Delos and she arrived just as Apollo was delivered into the world, aided by Artemis. A lunar Goddess, Artemis became henceforth the Goddess of midwives. In *The Nature of the Gods*, Cicero wrote that "Diana[3] [was] invoked at childbirth because children [were] born occasionally after seven, or usually after nine, lunar revolutions."[4]

Such is the story of Apollo's birth – a difficult, convoluted, painstaking one. Births matter a great deal to astrologers. In the horoscope the style and qualities of births and beginnings are indicated by the rising sign, the Ascendant. Apollo's Sun sign, undoubtedly, has to be Leo,[5] but given his very challenging birth, I would vouch for a Scorpio Ascendant – Scorpio births are often life-and-death struggles. Scorpio is a sign with plenty of *Sereti* – you do not mess with Scorpios. They sting and can hurt you. They are intense characters, not given to compromises, and they do not easily forget a wrong done them – think Charles Bronson, Scorpio Sun, in *Once Upon a Time in the West*. But they also carry medicine, usually delivered through a catharsis of some sort. One of the lesser-known attributes of Apollo was as the God of plagues. We find it mentioned in Homer. In the *Iliad*, Apollo is no friend of the Achaeans. Book 1 narrates how, after raiding a nearby town, Agamemnon and Achilles ended up fighting for the spoils of war in the person of a young woman, Chryseis. Agamemnon, as Supreme Chief, won the trophy, which sent Achilles sulking for a while – a central episode of the war. Chryseis, however, was the daughter of a priest of Apollo and the radiant God did not take the outrage lightly. He sent the plague raining down on the Achaean army gathered under the walls of Troy. Many souls made their way to Hades before Agamemnon, a rather ruthless man,[6] recanted under pressure from his agonizing men and released Chryseis.

Being associated with diseases Apollo was also a healer of diseases. He fathered Asclepius, the god of medicine, who was so skilled in his art that he once brought back a dead man to life – a hubristic feat that deprived Hades of a client and for which Zeus struck him dead with a thunderbolt.[7] Apollo was also the god of mice, the carriers of the plague. In fact, says Robert Graves, this is how he started: "The Greek God Apollo... seems to have begun as the Demon of a Mouse-fraternity in pre-Aryan totemistic Europe: he gradually rose in divine rank... until he became the patron of Music, Poetry and the Arts and finally, in some regions at least, ousted his father Zeus from the Sovereignty of the Universe."[8]

We are touching here on a very interesting pattern. From Mouse-Demon of a fraternity – Apollo Smintheus ('Mouse Apollo') was one of his earliest titles – to Sovereign God of the Universe, the role of Apollo has followed a gradual and profound evolution. This is seen for instance in his rulership over the Sun, which the God actually ripped from Helios, a Titan born of Gaia, whose sister was Selene, the Moon. What happened then to account for this shift? Things start to make sense when we acknowledge that our divinities are fluid images and that these images reflect our own consciousness. As Baring and Cashford point out in *The Myth of the Goddess*, "one way in which humans can apprehend and know their own being is by making it visible in the image of their goddesses and gods...

It [is] through such images... that consciousness tells its own story."[9] Consequently, adds Cashford, "the character of our divinities will change – must change – as our need of them is outgrown."[10] More than any other God, it seems, Apollo has been the recipient of our projections. Clothed in the garb of our increasing solar consciousness, he has radiated an ever-brighter light. And so has he become "the divine embodiment of reason and rationality."[11]

But as they say in the Kingdom of Fyfe, it takes a long spoon to sup with a Fyfer. To explore Apollo's image in some depth we may start from his birth on Delos. As soon as he was born Themis fed him nectar and ambrosia, the food of the Gods, while Hephaestus provided bow and arrow, his and his sister's trademark weapons. According to Graves,[12] Apollo's first deed was to go straight to Mount Parnassus to meet the serpent Python. Apollo had a score to settle with Python who had pursued his mother while she was pregnant. Severely wounded by Apollo's "solar rays",[13] the serpent fled to the Oracle of Mother Earth at Delphi. There, Apollo finally dispatched it near the sacred chasm. For Peter Kingsley, who has done much to reconstruct Apollo's image, "it's easy to assume that the Delphic myth of Apollo fighting the snake is a straightforward case of a battle between the opposites – of Apollo as a celestial god overcoming the powers of earth and darkness."[14] But Apollo, he adds, far from being antagonistic to snakes, had in fact a sacred connection with them:

> in ritual and in art snakes were sacred to him. Even in the case of the myth about the snake he fought and killed at Delphi, he didn't destroy it just to get it out of the way. On the contrary, its body was buried at the centre of his shrine. He killed it so he could absorb, appropriate, the prophetic powers that the snake represents.[15]

Alongside the patriarchal myth, another reading is then possible, one that speaks of transformation through a necessary sacrifice. And indeed Apollo the unforgiving god, the god of the plagues, took over rulership of Delphi and in time became known for his moderate and wise words: "Know thyself" and "Nothing in excess". Delphi was the most important site in Greece, the omphalos, the navel of the earth. The first ruler of the site was said to be Gaia, Mother Earth. The Python was Gaia's child, but in another version, Gaia passed the stewardship to her daughter Themis who willingly handed it over to Apollo and continued to work alongside him. Themis is sometimes said to be the first Pythia. For scholar Sarah Iles Johnston, however, there is no evidence that Gaia was the primary divinity worshipped at Delphi: "stories about Gaia's early ownership of Delphi must be read instead as one variation of a theme that runs throughout Greek myth, according to which female power initially held sway in the universe but gradually was replaced by male power."[16] A theme we are quite familiar with. Kerényi is of a different opinion though: "on the strength of the tradition and of archaeological findings, the existence of a pre-Apollonian earth oracle at Delphi may be taken for certain."[17] Iles Johnston adds that "the stories probably also reflect the persistent belief that sources of prophetic power dwelt in or below the earth"[18] – in the womb of the earth, as it were, at which point we may note that 'Delphi' is etymologically linked to the word 'womb'. Be that as it may, Apollo became the god of prophecy, an art that he was sometimes supposed to have learnt from the god Pan. For Iles Johnston, oracular consultation was a multi-stage operation:

First, Zeus determined his will. Next, Apollo articulated that will and the ways it impinged upon humans. Then, the Pythia spoke on Apollo's behalf, following which her listeners scrutinized and decoded Apollo's words. Finally, the listeners acted, making whatever arrangements were necessary to carry out what they took to be Apollo's advice.[19]

Contrary to our modern understanding of oracles, oracular consultation was rarely about asking, "What will happen?", but rather about seeking an answer to the much more important question, "Are the gods willing?" This is the original meaning of *divination*, from *deva* – the Sanskrit word for 'deity'. To divine is to probe divine will. The Pythia was the instrument of that will: seated on a tripod placed over the sacred chasm in the ground, she was becoming enthused, filled with the God who was taking possession of her. "At Delphi, and apparently at most of his oracles," says ER Dodds,

> Apollo relied on 'enthusiasm' in its original and literal sense. The Pythia became *entheos, plena deo*: the god entered into her and used her vocal organs as if they were his own...: that is why Apollo's Delphic utterances are always couched in the first person, never in the third.[20]

The statements uttered by Apollo were notoriously opaque and confusing. I find them extremely poetic though. Maybe this is because Poetry and Prophecy, when divinely inspired, meet in madness. "The divine madness was subdivided into four kinds," said Plato in the *Phaedrus*, "prophetic, initiatory, poetic, erotic, having four gods presiding over them; the first was the inspiration of Apollo, the second that of Dionysus, the third that of the Muses, the fourth that of Aphrodite and Eros."[21] Primacy was assigned to the latter.

In 480 BCE, the Greeks, preparing to meet the Persian King Xerxes and his powerful army at the battle of Thermopylae, consulted the Pythia, who ominously uttered:

> Now your statues are standing and pouring sweat
> They shiver with dread

The black blood drips from the highest rooftops
They have seen the necessity of evil
Get out, get out of my sanctum and drown your spirit in woe.[22]

This did not sound good and the Greeks duly lost to the Persians. No ambiguity there but a powerful poetic imagery. In *Battlestar Galactica*, we encounter the hybrids, amongst the most intriguing characters of the series. The hybrids are a special breed of Cylons, constantly immersed in a bath of creamy water, where conduits and connectors seem somehow part of their bodies. Sort of living computers, their function is to manage the basestar which they inhabit. One of their fascinating features is to constantly voice incomprehensible sentences, some related to the functions of the basestar, and some resembling prophetic utterances. Some Cylons believe that the hybrids can hear or understand the voice of God, some others that they have seen the place between life and death and have gone mad as a result. Anyway, I cannot help but think of the hybrid as a Pythia on steroids:

The excited state decays by vibrational relaxation into the first excited singlet state. Yes, yes and merrily we go. Reduce atmospheric nitrogen by 0.03%. It is not much consolation that society will pick up the bits, leaving us at eight modern where punishment, rather than interdiction, is paramount. Please, cut the fuse. They will not harm their own. End of line. Rise and measure the temple of the five. Transformation is the goal. They will not harm their own. Data-font synchronization complete.[23]

There my mind wanders to Surrealism, with which I became totally infatuated in my twenties. At the time it felt like drinking from a new, refreshing spring opened up by Pegasus' hoof. In its poetical expression, surrealism is assuredly very close to madness. Witness this extract from *Soluble Fish*, written by André Breton in 1924:

> HELEN: The window is open. The flowers smell sweet. Today's champagne, a glass of which is bubbling in my ear, makes my head swim. The cruelty of the day molds my perfect forms.
> SATAN: Can you see the Isle Saint-Louis above these Ladies and Gentlemen? That's where the poet's little room is.
> HELEN: Really?
> SATAN: Every day he was visited by waterfalls, the purple waterfall that would have liked to sleep and the white waterfall that came through the roof like a sleep-walker.
> LUCIE: It was I who was the white waterfall.[24]

While one may object that we are here far from divinely inspired madness, we may retort: how does the modern mind define 'divine'? Define 'madness'? In *The Manifesto of Surrealism*, Breton asserts that "it is true of Surrealist images as it is of opium images that man does not evoke them; rather they come to him spontaneously, despotically. He cannot chase them away; for the will is powerless now and no longer controls the faculties."[25] This sounds like a proper form of possession – by the unconscious, the new unruly God of psychology. I often wonder what the Ancients would have made of the modern forms of art. What would Sappho have said to TS Eliot and Praxiteles to Barbara Hepworth? When French poet Reverdy writes: "Day unfolded like a white tablecloth," are we not quite close to Homer's "Rosy-fingered dawn" – or are we thirty centuries away? I find it striking that as the world moved towards increased literalism and away from the symbol, art followed an opposite curve. Impressionism as a counterpoint to Industrialization, and Surrealism in reaction to Modernity. But then it feels like at some point in the 20th century, Art gave up the fight and surrendered, falling head-on on to postmodernist meaninglessness. Andy Warhol comes to mind, and so do Saatchi's YBAs. But we may blame Duchamp for starting it all with his urinal. Nowadays, artworks come with an instruction manual, just like washing machines. One has to enter the mind of the artist to understand what is meant by the work. Forget soul, the tools of choice are the intellect and reason, a reflection of our bloated rationalism. No more flickering candles but buzzing neon lights. For the most part, it leaves us dry and vaguely nauseous. Nay, let me rephrase that, for I have no right to speak for everyone: for the most part, it leaves *me* dry and vaguely nauseous. I readily concede, however, that it may be the whole point of the exercise.

Apollo was the god of music and poetry, although while he was indeed an unrivalled musician on his lyre, he deputed poetry to the Muses. There is here an interesting evolution for by the time of Nietzsche, Apollo was no longer associated with music – Dionysus was: "Intelligence is often the enemy of poetry," said Garcia Lorca,

> because it limits too much, and it elevates the poet to a sharp-edged throne where he forgets that ants could eat him or that a great arsenic lobster could fall suddenly on his head – things against which muses that live in monocles and in the lukewarm, lacquered roses of tiny salons are quite helpless.[26]

Has Apollo become too intelligent? Too far removed from his art, too comfy in the lacquered roses of tiny salons? Too self-aware of his own light to ever consider the great arsenic lobster precariously hanging over his mighty crown of laurels? Hillman certainly thought so. But Apollo, peace be on him, is in no way responsible. *We* are, coaxing his muses into wearing monocles. By the end of the 19th century, after centuries of Sun dominance, with Reason and Rationality in its wake, Apollo's image had been thoroughly revised. In a famous essay, *The Birth of Tragedy*, published when he was only 28, Nietzsche described Apollo as "that measured limitation, that freedom from the wilder emotions, that philosophical calmness of the sculptor-god. His eye must be 'sunlike'."[27] A razor-like God. Apollo, the shaper of forms, was contrasted by Nietzsche with Dionysus, the dissolver, and in this antithesis resided creativity:

It is in connection with Apollo and Dionysus, the two art-deities of the Greeks, that we learn that there existed in the Grecian world a wide antithesis, in origins and aims, between the art of the shaper, the Apollonian, and the non-plastic art of music, that of Dionysus… continually inciting each other to new and more powerful births, to perpetuate in them the strife of this antithesis, which is but seemingly bridged over by their mutual term 'Art'.[28]

Isn't it symbolically significant that in the summer of 1969, at the end of a decade of immense creativity, we find two major events perfectly encapsulating the Apollonian and Dionysian energies, two events that each in their own way have contributed to define the 20th century? On 20 July, the Apollo Moon landing, a "sunlike" endeavour, a technological prowess, the triumph of hardnosed, no-nonsense, cool Science – that "freedom from the wilder emotions". But a science subservient to a higher, mundane goal: the space race used as a pawn in a geopolitical game – throwing the gauntlet into orbit at the old, and cold, enemy. Barely a month later, on August 15, the Woodstock music festival, a huge release of Dionysian energy, a human tidal wave dissolving in mud, music and intoxication: "Woodstock was a spark of beauty," said Joni Mitchell, "where half a million kids saw that they were part of a greater organism." Here we meet the archetype of Neptune, the planet of dissolution, idealism and surrender, a planet closely linked to Dionysus and, as the ruler of Pisces, to Christ. A planet also connected to Jimi

Hendrix: Neptune was the most elevated planet in his birth chart, at 1-degree Libra. Hendrix, the arch figure of the Sixties, who in the pale dawn of Woodstock closed the proceedings by deconstructing the American anthem, most kids already gone, not yet aware that they had made history! In Hendrix's chart, Neptune played its role as the planet of inebriation, a Dionysian theme as well. Hendrix also had Sun, Mercury and Venus in fiery Sagittarius, the sign of expansion, and the temptation was too great. Higher and higher he went on the eagle's back until the final combustion, a modern Icarus in Wonderland flying too close to the Sun, eventually crashing down to the sea, intoxicated with barbiturates, choking on his own vomit – powerful Neptunian signatures, all. The whole rock 'n' roll scene of the late sixties can in fact be seen as embodying the Dionysus archetype. Is there a better faun than Mick Jagger, his lascivious mouth exuding the grunts and groans of a roaming predator? No doubt Dionysus would have thoroughly enjoyed the Fillmore West in San Francisco, where Bill Graham was mixing night after night a brew of cosmic proportions. The leprechauns were dipping their crooked fingers in the cauldron and to hell with work, family and country!

I have always had a particular fondness for the Sixties. Maybe because this is the decade of my birth. I like to think that there is more to it than that though. More than any other decade in the 20th century the Sixties, I feel, shine and radiate a very particular light. I cannot help but see this period as the hinge on which our world revolves. A door opened then, a glimpse of what might be. It is as if all the gods of Olympus decided to have a hell of a party, descended on Earth, shook their booty and went raving mad. All the archetypes were activated, intensified, infecting everything with their red saliva, a rainbow of Day-Glo colours, a merry-go-round of pulsating energy. I am nostalgic of the Sixties without having lived through them. I am sure the old hands would say that I am nostalgic precisely because I did not live through them! Maybe. But to me, it seemed like things were done with meaning. Even meaninglessness was meaningful. There was an attitude, a nailing of the colours to the mast, a refusal of the status quo. One still believed back then, maybe foolishly but resolutely, and with style! On the streets wandered "angelheaded hipsters" – right, Mr Ginsberg? – with diamond dust in their hair, "burning for the ancient heavenly connection to the starry dynamo in the machinery of night."[29] I know the usual criticisms – that it was a middle-class 'revolution' (give me one that's not!), spoilt kids who had it easy, hedonists looking for an easy lay, escapists on the wings of white rabbits, and a huge con for women. Points taken on all of them, especially the last one. It was not cool to refuse to sleep around – and who was cooking in the communes? Venus was central to the party and quickly became disenchanted. And when Venus is unhappy, she manifests herself in pretty grim ways – witness the free health clinics all over Haight-Ashbury to handle the spread of *vene*real infections, those archetypal diseases of a Venus run amok. Maybe what I am talking about is the initial spark, before everything became corrupted by the merchants of death, by the devitalizing gaze of the squares, by the crushing weight of the media, by the bone-sucking commoditization of the imagination. Everything becoming an object to be turned into mugs and fridge magnets. Che Guevara on denim caps and 'Peace and Love' on trendy shirts, produced in Vietnam for a few pennies, sold on the high street for £50. Nothing sacred, everything for grabs. "What sphinx of cement and aluminum bashed open their skulls and ate up their brains and imagination? Moloch! Moloch!" raged Ginsberg a decade before.[30] Already by October 1967 the early residents of Haight-Ashbury had understood where this was all going. "Moloch whose mind is pure machinery!" On 6 October, they organised the Death of the Hippie, a funerary procession in town. The dream was over, before it was repackaged for a wider audience. "Moloch whose eyes are a thousand blind windows!"

So there – it lasted 18 months, two years at the most, this pure and idealist impulse, before the hungry dogs took over. This is the spark I am nostalgic about, not the poignant rigmarole that followed. Not that it did not have its beautiful moments too though, and Woodstock, certainly, was one – as were countless other radiant sparkles, before Altamont and Charles Manson drove the final nail in the proverbial coffin.

But this juggernaut, the unruly gods vying for attention, the Olympian brawl – that is quite something! The whole decade was being sensitized by the conjunction of Uranus and Pluto, like an electric current throbbing in the veins – a conjunction that happens less than once a century. In *Cosmos and Psyche*, his extensive study of astrological collective cycles, philosopher Richard Tarnas associates this conjunction with the awakenings of the Dionysian:

> The Plutonic-Dionysian archetype, associated with the planet Pluto, [intensifies] and [empowers] on a massive scale the Prometheus archetype of rebellion and freedom, creativity, innovation, and sudden radical change, all associated with the planet Uranus.[31]

For indeed it was not all roses and frankincense. Mars had joined the fray, craving blood in Vietnam, Paris or Prague, rioting in LA and on the campuses, birthing the Cultural Revolution in China. Apollo and Dionysus were going head to head, the forces of change and dissolution versus the existing structures of science and politics. In the parties in town, one question was repeatedly hushed over canapés and spiked drinks: are you rather Moon-landing or Woodstock? So many doors had been opened that one might say that the whole palace had been blown apart. And Apollo, this delightful God of Creativity and the Arts, had in spite of himself almost become Saturnian – stern, rigid, conservative. A mockery of himself, a caricature.

Away with the lyre, then. By the late 19th century Apollo had become the Sculptor, the patron of Doric art, and had lost dominion over music – which, according to Nietzsche, he never really possessed in the first place anyway:

> If music, as it would seem, was previously known as an Apollonian art, it was, strictly speaking, only as the wave-beat of rhythm, the formative power of which was developed to the representation of Apollonian conditions. The music of Apollo was Doric architectonics in tones, but in merely suggested tones, such as those of the cithara. The very element which forms the essence of Dionysian music (and hence of music in general) is carefully excluded as un-Apollonian; namely, the thrilling power of the tone, the uniform stream of the melos, and the thoroughly incomparable world of harmony.[32]

There was a time, though, when Apollo did not take it lightly to be defied on his musical abilities. A famous story saw him challenged by Marsyas, a Phrygian Satyr, to a musical contest. To cut a story short, Marsyas eventually lost and Apollo showed his ruthless streak by having Marsyas tied to a tree and flayed alive. In Greece, hubris always called for nemesis. Interestingly, Marsyas was said to be connected to Dionysus, playing the flute in the retinue of the wandering god. The Apollo/Dionysus polarity is truly immensely rich, playing out in many varied ways. Its richness may come from its archetypal quality, for it brings us back to Sun and Moon – Apollo the solar god, and Dionysus the dead and resurrected lunar god. It may also bring us back to Christ, for in Christ, Apollo and Dionysus may be said to meet – Christ as the Apollonian light of the world, and Christ as the Dionysian sacrificed Son of God. Mithras never stood a chance, really – too uni-dimensionally solar.

In *The Four Faces of the Universe*, philosopher Robert Kleinman explored Beethoven's *Ninth Symphony* from a cosmological angle and used this very polarity to contrast the second and the third movements:

> SECOND MOVEMENT (Scherzo): The play of wild Dionysian energies in the throes of creation.
> THIRD MOVEMENT (Adagio): Serene Apollonian reflection on what has been accomplished, followed by a disturbing sense that something is still lacking.[33]

Apollo and Dionysus, the half-brothers, were said to share Delphi, with Dionysus in charge when Apollo retired to the Hyperborean land for the three winter months. Hyperborea was a mythical land, often associated with Britain, sometimes referred to as the land beyond the northern wind, or "the land of long-lived men, living in a realm of eternal spring in the far north."[34] Apollo's mother was said to come from Hyperborea, and the Hyperboreans were totally devoted to Apollo and Artemis. But as ever, versions of the same myth did not necessarily match. For Robert Graves, quoting the historian Hecataeus (6th century BCE), "Apollo visits the island once in a course of nineteen years, in which period the stars complete their revolutions... During the season of his appearance the God plays upon the harp and dances every night, from the vernal equinox until the rising of the Pleiades, pleased with his own successes."[35]

From the vernal equinox to the rising of the Pleiades: for Hecataeus it is in summer, then, that Apollo visits the island, not winter. And that, only every nineteen years. Maybe we should try not to be too caught up in splitting hair. Even the sharing of Delphi was not unequivocally understood: according to the Greek poet and scholar Callimachus (3rd century BCE), "when the Titans had torn apart Dionysus, they gave the limbs to his brother, Apollo, having thrown them into a kettle, but he preserved them close to the tripod."[36] A very different kind of sharing, indeed.

For German scholar Walter Burkert, underlying the myth of the sharing of Delphi lies

> a polarity in which the contrary elements determine each other. It comprises savagery versus clarity, lack of inhibition versus awareness of limitations, female versus male, proximity to death versus affirmation of life: this is the circular course that sacrificial ritual charts again and again, renewing life by encountering death.[37]

I wonder what Dionysus made of the 'Nothing in excess' precept of his brother. Did he cringe every time he saw it? Was he tempted to tag it when Apollo was away, like a mischievous child, with a crazy grin on his distorted face – an irreverent Delphic Banksy?

But I am being flippant. Dionysus was much more than a misbehaved child. The Greeks had two words for life: *zoe* and *bios*. The latter referred to individual life, the life of a specific organism as distinct from other forms of life, a life that inevitably ends at some point. The former was Life in general without further characterization. As Kerényi puts it, "if I may employ an image for the relationship between them… *zoe* is the thread upon which every individual *bios* is strung like a bead, and which, in contrast to *bios*, can be conceived of only as *endless*."[38] *Zoe* is mirrored in the rhythms of the Moon, dying every month in an endless cycle of life. *Bios* is what the heroic solar ego associates with, hence his terror of death seen as a 'dead end' (sic). Kerényi saw in Dionysus an archetypal image of indestructible life, a divine symbol of *zoe*.

Coming from a more psychological angle, Jungian Edward Whitmont saw the Apollo/Dionysus conflict played out in the human psyche as a collective "thou shalt" repressing individual urges:

> Discipline and obedience to rules require the repression

of spontaneous needs and urges... Natural spontaneity, sexuality, the desires of the flesh, woman and the Feminine, dance and play, all become the powers of the adversary, Dionysus made into the Devil. They are dreaded and repressed.[39]

In saying this, Whitmont was not advocating an uncontrolled release of Dionysian energies, which would be as destructive as their repression. Rather he was insisting on the importance of ritual to express, honour and channel these energies in a safe way. For indeed how can we reconcile the two sides and give each one its due? For psychologist Helen Luke, the Greeks still knew how to tread the middle ground – just. In Aeschylus' time, she says, a split was appearing away from the feminine values of nature, a split that Aeschylus explored in his play the *Eumenides*. Orestes is pursued by the Furies for his matricide, an unforgivable crime, and implores Apollo's help. Apollo and the Furies – Reason versus Instinct. Apollo advises Orestes to seek advice from Athena who, says Luke, "was for the ancient Greeks the symbol of masculine courage and understanding united to feminine feeling and instinctive wisdom."[40] A proper solar goddess. Apollo had rejected and condemned the Furies as plain evil but Athena goes deeper:

It is as though Athena penetrates to the inner meaning of the Delphic oracle. It is true that man must free himself from the devouring mother who murders the values of consciousness, but not by means of another murder. Real freedom can only come through accepting inwardly the furies of instinct without loss of conscious values, thus redeeming them.[41]

It is about somehow making space and accommodating the exiled Feminine without being submersed by Her power. It is about consciously and compassionately welcoming back Medusa in our midst, and there is no doubt that it is a momentous task – but arguably, it may be the most urgent task at hand.

Three Powers were at play in Greece, says French scholar Jean-Pierre Vernant, that could be seen to represent instinct as a form of 'alterity' or 'otherness' – three deities "whose cults contain masks, either votive offerings or objects carried by the celebrants,"[42] as if to hide their unacceptable face: Dionysus, Artemis and Medusa. I find it puzzling that two of them were directly associated with Apollo, as if Apollo was calling for contrast with the Other, for polarity with the Outcast.

While Medusa stands for pure, terrifying chaos, says Vernant, and Artemis is the Wild One,

> with Dionysus, the music changes. At the heart itself of life on this earth, alterity is a sudden intrusion of that which alienates us from daily existence, from the normal course of things, from ourselves: disguise, masquerade, play, theatre, and finally, trance and ecstatic delirium. Dionysus teaches or compels us to become other than what we ordinarily are, to experience in this life here below the sensation of escape towards a disconcerting strangeness.[43]

This 'otherness' created a very tight bond between Artemis and Dionysus – he was, after Apollo, the god most associated with her: "it is not surprising," says Sorita d'Este, "that there should have been a bond between Artemis, as a goddess of the wilds worshipped by women, and Dionysus, a god of the wilds worshipped by women."[44] As Lady of the Wild many animals were sacred to Artemis, particularly the bear. But the boar was among her iconic animals too. At Ephesus, said Herodotus, Artemis' temple had been founded on the spot where a boar had been killed, fulfilling the words of the oracle that "a *bird and a boar will show you the way*."[45]

Although no Artemis I, too, have a particular connection to

the wild boar. I attended once a shamanic workshop in London. At the end of the weekend our task was, working in pairs, to journey for our fellow student and encounter his or her power animal. When the woman with whom I was working journeyed for me I was rather hopeful that she would meet a stag, a wolf or an eagle, one of those noble animals with whom everyone wants to work. But to my great disappointment, she had met a boar. The Boar, then, was my power animal. I was deflated but I quickly realized that it was indeed the perfect guide for me. The boar is close to the ground, compact, fierce, determined, earthy, raw – qualities that are not necessarily mine to start with, and that are powerful allies to call upon when needed. Spirit does not give you what you want, it gives you what you need, and too bad if it ruffles the peacock feathers of your little ego. I did invoke the Boar a great deal when I had to face trial in Paris, and he was of invaluable support.

I remember an episode a few years before the shamanic weekend. It was on the second day of my stay in police custody. I had been interrogated all morning and lunchtime, and in the early part of the afternoon I was placed in a cell, waiting for the police to sort out their notes before bringing me back in for further questioning. I was sitting in my cell when the door opened and a man was brought in. Of Italian origin, in his late fifties, he looked like an innocuous jolly granddad and, in true Italian style, was relentlessly talkative. It was good to have a companion, especially a lively one, full of anecdotes. He explained that he was being suspected – totally unfairly, he emphasized – of being a bit too creative with the accounting of his small building firm. Anyway, I do not know how we got there but he started to talk profusely about wild boars, and how there were 22 million of them roaming the French countryside, causing extensive damage, especially in the South. He sure seemed to know a whole lot about boars. I am sure that we talked about many other things too – we must have been together a good hour – but the only conversation that sticks to this day is his talk about boars. I remember how good it felt, how comforting it had been in this time of darkness. Looking at it from today's vantage point I can see that my ally the Boar – whom at the time I did not even know qualified as such – had somehow made its way into my cell to give me support and strength.

In truth, we are terribly unfair to Apollo, straightjacketing him into a role that he never called for. For Peter Kingsley, Apollo was initially the god of the Greek shamans, the god of incubation, and there was, too, an otherness about him, which is why he may be so closely associated with Dionysus: "he was a god of ecstasy, trance, cataleptic states – of states that takes you somewhere. There was a single word in Greek to express this; it meant 'taken by Apollo'."[46] But Kingsley is keen to emphasize that Apollonian and Dionysian ecstasies were very different in nature:

> There was nothing wild or disturbing about [Apollonian ecstasy]. It was intensely private, for the individual and the individual alone. And it happened in such stillness that anyone else might hardly notice it or could easily mistake it for something else. But in this total stillness there was total freedom at another level.[47]

At some point, however, our image of Apollo changed. Partner in crime with the Sun, he bore the brunt of our solar projections and something important was lost in the process: "when people started trying to make Apollo reasonable, philosophically acceptable, they were simply looking at the surface and avoiding what's underneath."[48] For Kingsley, what was 'underneath' was Apollo's relation to Persephone and to the Underworld: "Apollo is the god of healing but he is also deadly. The queen of the dead is the embodiment of death; and yet it was said that the touch of her hand is healing. As their own opposites they exchanged roles with each other – and with themselves."[49] In this relation we may recognise the image of the midnight Sun, the Sun of darkness symbolic of "an illumination that comes from below".[50]

In this long process of revisionism, it was not only music that was ripped from Apollo's domain but poetry as well. Such a figure as Blake would equate Apollo with Urizen ('your reason') and by extension to Satan: "I have conversed with the Spiritual Sun – I saw him on Primrose Hill. He said, 'Do you take me for the Greek God Apollo?' 'No,' I said, 'that [pointing at the sky], that is the Greek Apollo – He is Satan.'"

Apollo now carries our shadow. Referring back to Aeschylus' *Eumenides*, Helen Luke stresses that

[Athena] knows that, if civilized justice and reason become identified with the Apollonian values as the only good, the destructiveness of the ancient mother goddesses will erupt in untold evil for mankind. The 'murder' of the instinctive and the irrational by civilised consciousness will result in the final poisoning of everything.[51]

We have now reached that stage. Apollonian consciousness has become a byword for patriarchy, as feminist social writer Camille Paglia curtly asserts: "The Apollonian is harsh and phobic, coldly cutting himself off from nature by its superhuman purity. I shall argue that western personality and western achievement are, for better or worse, largely Apollonian."[52] "The Apollonian has taken us to the stars,"[53] concedes Paglia but this, for Ginette Paris, is the inevitable outcome of the Apollo archetype: "To leave the Earth, the Mother, and approach the Sun is the Apollonian ideal, Apollo never being high enough or far enough."[54] By looking up, however, we have forgotten to look down:

We have allowed our planet, our collective household, to deteriorate, the ashes scattered, and everything in disrepair.

Why should we do repairs when so many of us live as if we were temporary 'tenants' of this little planet, which we may well leave one day, as soon as our scientific heroes give us the signal?[55]

Popular culture, undoubtedly, supports that stance. On 3 June 1957, *LIFE Magazine* reported about the American *Vanguard* project, whose aim was to launch the first artificial satellite into Earth orbit, in the following terms: "If the rocket does launch its moon into space, it will mark man's first step in breaking the bonds that chain him to his native planet."[56] *The bonds that chain him to his native planet* – why indeed, as Paris says, pay any attention to a jailhouse, a confinement cell? We might just as well trash the place and move on.

This is where *Battlestar Galactica* comes in, again – for me, at least. I think that it is only at the end of the story that I understood why I was so captivated by it. The series go over four seasons and each episode is packed tight, intense, relentless. The writers take no prisoners and ask of the viewer a total commitment. The sense of dread is overwhelming, and so is the increasing claustrophobia. Towards the end, the vessel is dying, beyond repair – and the characters, too, seem beyond repair. We are suffocating with them in the enveloping darkness, we are breathing the same vitiated air. The frontier between humans and Cylons has become totally blurred, and so is the polarity between life and death, between truth and lies, between spirit and matter. But after much travelling and hardship, after having gone through untold suffering and countless challenges, they find their way home. The camera follows the creaking battleship, which is about to crumble and die of old age. We see the spaceship glide over what looks like familiar ground, a silvery and dusty surface – the Moon. And beyond the Moon, there She is – radiant, glorious, glittering, a shining diamond. I find this scene incredibly moving. *Battlestar Galactica* is a modern *Odyssey*. A homecoming. Ulysses returns to Penelope, Humans return to Earth. Significantly, when the band of rags-and-bones exiles land on the surface, what they find is Earth at the dawn of humanity. A pristine land, a story to be written, a new beginning, the chance to start afresh. Commander Adama becomes Adam, the first man. And yet here we are today and the cycle seems to have repeated itself. Which is where the Apollo mission has so much to teach us, for the lunar landing, too, tells of a homecoming. It gifts us with a new vision of Home, symbolised by the *Earthrise* picture taken by *Apollo 8*. This new vista is compelling – it speaks of the necessity to renew our bond with Earth and, beyond, with a lunar consciousness that has been obliterated for too long by Apollo's pathological light. We are not chained to our native planet. We are One with Her.

More than anybody, James Hillman has stood as a ruthless and uncompromising critic of Apollonian consciousness and, echoing Whitmont, was wont to charge Apollo with many of the ills of this world, and in particular our fraught relation with the Feminine:

> [The Apollonian] belongs to youth, it kills from a distance (its distance kills), and, keeping the scientific cut of objectivity, it never merges with or 'marries' its material. It is a structure of consciousness that has an estranged relation with the feminine, which we have taken to mean 'the abysmal side of bodily man with his animal passions and instinctual nature', and 'matter' in general.[57]

But Hillman was also careful to distinguish between archetype and gender: "the Apollonic fantasy… is not exclusively male… As an archetypal structure it is independent of the gender of the person through whom it works, so that the integration of the feminine is a concern pertaining not only to men but to women as well."[58]

In Dionysus Hillman saw an archetypal figure that could heal the split generated by an excessive Apollonic consciousness: "though he is male, and phallic, there is no misogyny in this structure of consciousness because it is not divided from its own femininity."[59] Dionysus, the bisexual god – an androgynous consciousness, where male and female are primordially united. A different approach to Nietzsche then, who saw in the neat polarity of the two brothers the spark of creativity. For Hillman, Apollo has failed and Dionysus comes to the rescue, a figure reconciling in himself masculine and feminine. With the most sincere respect due to Hillman I am, however, a bit reluctant to deny Apollo a fair trial. Like a Dorian Gray in reverse, the Sun-God now walks amongst us carrying a distorted projection entirely of our making while his true, luminous and deep nature is hidden away in the attic. We threw his soul to the hungry dogs of our "irrational rationality", as psychologist Erich Fromm called it. As for Dionysus he is, despite his bisexual nature, male and I feel ambivalent about using another male figure, however androgynous it may be, as the ideal new archetype. I am infinitely more comfortable with honouring Apollo the solar masculine and Dionysus the lunar masculine, on the one hand, and Athena the solar feminine and Hekate, whom we will encounter later on, the lunar feminine, on the other – all four of them equal under the Sun and under the Moon.

And then, the dance.

Chapter 4

The Demotion of the Moon

From Goddess to Real Estate

Are you going to tell me [the Moon] is a dead lump?... When we describe the Moon as dead, we are describing the deadness in ourselves.
– DH Lawrence

In a perfectly orchestrated double act, Copernicus and Galileo drove the final stake through the heart of the Moon in the Renaissance period. Copernicus' *De Revolutionibus* was published in 1543, on the very day of his death, and although it took decades to shake off the old Ptolemaic paradigm, the world was no longer the same. In a supreme act of Imagination, Copernicus had left the Earth and placed himself on the Sun – and theorized how the movements of the planets would look like from that vantage point. All of a sudden, a lot of things started to make sense, in particular the retrograde motions of the planets that had always been a very sticky point in the Ptolemaic model. But of course, by standing outside the Earth and looking back on her, the Earth herself was affected in a major way. She was not the central focus anymore. She was moving around the Sun, like all the other planets. While one may argue that the Copernican Revolution essentially turned the relation between Earth and Sun on its head, leaving the Moon relatively unaffected, the mere promotion of the Sun further contributed to the demotion of the Moon. For science had suddenly provided scintillating evidence to what had been known for three thousand years: when one raised one's eyes to the Heavens the Sun ruled supreme, and the Moon was merely subservient to the Earth. As philosopher of science Thomas Kuhn said in *The Copernican Revolution*:

> In the middle of all sits the Sun enthroned. In this most beautiful temple could we place this luminary in any better position from which he can illuminate the whole at once? He is rightly called the Lamp, the Mind, the Ruler of the Universe... So the Sun sits as upon a royal throne ruling his children the planets which circle around him.[1]

If the Moon managed to retain a halo of mystery after Copernicus, the inquisitive eye of Galileo's telescope made sure this faint light would evaporate once and for all:

> I have been led to that opinion... that I feel sure that the surface of the Moon is not perfectly smooth, free from inequalities and exactly spherical, as a large school of philosophers consider... but that, on the contrary, it is full of inequalities, uneven, full of hollows and protuberances, just like the Earth itself.[2]

Not only had Galileo found evidence of the rugged and not-so-special nature of our Moon, he had also uncovered the presence of *four* Moons around Jupiter. No doubt other Moons elsewhere were waiting to be discovered in the known universe. The Moon, this Moon in our sky that had enchanted us for so long, was no deity, no Goddess – it was a mere lump of rocks, and it was not even exclusive to Earth. Centuries later, *Apollo 8*'s Jim Lovell was expressing a similar sentiment: "[The Moon] is ugly... I am kind of curious how all the songwriters can refer to it in such romantic terms."

The Earth moved around the Sun, not the other way around, and the Moon was an unremarkable sphere that did not seem to have any utility. We had been conned. Not only were our senses unreliable but, as Hannah Arendt pointed out in *The Human*

Condition, even rational truth was questioned:

> If the human eye can betray man to the extent that so many
> generations of men were deceived into believing that the sun
> turns around the earth, then the metaphor of the eyes of the
> mind cannot possibly hold any longer; it was based... on
> an ultimate trust on bodily vision. If Being and Appearance
> part company forever, and this – as Marx once remarked –
> is indeed the basic assumption of all modern science, then
> there is nothing left to be taken upon faith; everything must
> be doubted.[3]

In this dispiriting realization we hear the cries of modernity
being birthed, at the heart of which, says Arendt, was *doubt*. The
whole Cartesian inquiry started with doubt, desperately trying
to answer the "two interconnected nightmares – that everything
is a dream and there is no reality and that not God but an evil
spirit rules the world and mocks man."[4] And if we thought
that Newtonian physics, in its ability to predict and control
the material world, was at least providing us with a modicum
of solidity, think again. Enter quantum physics, all mind-
bogglingly smoke and mirrors, one fundamental tenet of which
is Heisenberg's *Uncertainty Principle*. There, it is said: we are
on very shaky ground. Faith, too, of course, had become highly
suspect. When the Church, stuck into dogma and hardened
by the Reformation, forced Galileo to recant, it did itself a
huge disservice that is still being branded today as proof of its
irrelevance and obscurantism. It may be convincingly argued
that humanity had to go through a period of doubt, iconoclasm
and rejection so as to stand outside faith and belief – which all
too often decomposed into superstition, delusion and abuse. But
has the pendulum swung too far in the other direction? Belief
in scientific truth is just another belief, that is, a psychological
projection, as Edinger points out:

I don't doubt that someday all our present cosmological notions will be recognized as myths, just as Ptolemy's notions are now recognized as myth. What we now think is science will, in the future, very likely be seen as myth – in other words as a psychological projection.[5]

In 2009, psychiatrist Iain McGilchrist published *The Master and His Emissary*, a groundbreaking study of the left and right hemispheres of the brain. To speak of left brain and right brain has become a bit of a *cliché*, which is unfortunate because it is a very relevant epistemological tool. In his book, rather than focusing on the 'What', McGilchrist looks into the 'How'. In other words, he stresses, the difference between left and right is not so much a question of differentiated functions than that of similar functions performed in a different way. Take for instance a bird foraging on open space. This bird wants food but is also, potentially, food. A two-tier operation is therefore at play. While one part of the brain will narrowly focus attention so as to find the seeds and tiny specks of food on the ground, another part will stay alert to the wider environment, keeping a large angle of vision so as to be on the lookout for potential predators. The left brain caters to the former operation, the right brain to the latter – "two widely differing ways," says McGilchrist, "of attending to the world."[6] But they are not perfectly equal partners: for its flexibility, its breadth of vision, its problem-solving capabilities, its creativity, its capacity to think in, and form, images, its wholeness, its holistic approach, the right has primacy. The left brain must be in service to the right, insists McGilchrist, if only because while the right can accommodate the left, the reverse is not true:

> The more flexible style of the right hemisphere is evidenced not just in its own preferences, but also at the 'meta' level, in the fact that it can also use the left hemisphere's preferred

style, whereas the left hemisphere cannot use the right hemisphere's.[7]

Therein comes the metaphor of the Master (the right brain) and his emissary (the left brain). At some point, says McGilchrist, the emissary staged a coup and took control of the Master. Why and how, one might ask, did the right brain relinquish its position? Because, McGilchrist continues, the Right depends on the Left to realize its higher purpose – a key point that is worth quoting at length:

> The right hemisphere, the one that believes, but does not know, has to depend on the other, the left hemisphere, that knows, but doesn't believe. It is as though a power that has an infinite, and therefore intrinsically uncertain, potential Being needs nonetheless to submit to be delimited – needs stasis, certainty, fixity – in order to Be. The greater purpose demands the submission. The Master needs to trust, to believe in, his emissary, knowing all the while that the trust may be abused. The emissary knows, but knows wrongly, that he is invulnerable. If the relationship holds, they are invincible; but if it is abused, it is not just the Master that suffers, but both of them, since the emissary owes his existence to the Master.[8]

Left versus Right, Reason versus Revelation, Sun versus Moon, Masculine versus Feminine – the temptation is great to fall into a very polarized view. Of course this very polarity is typically left brain – seeing things in black or white shades, an either/or mentality. But if one can see the polarity playing out into a larger, unified whole, as illustrated with the *taijitu*, the yin/yang symbol, confrontation gives way to dance.

In her cycle, the Moon symbolises this larger whole, a whole that encompasses dark and light, a both/and function that is also comfortable with either/or because it is wider in scope – while the opposite is not true. In that sense the Right hemisphere makes a good analogy with the Moon. Can we somehow see a thread slowly emerging? Left is as necessary to Right as Right is to Left, yet Left has to be ultimately in service to Right. Must the Sun be in service to the Moon, then? Well, truly, He may be said to be – by fecundating Her with His light. This will be the object of a further chapter, where the intermediary role of the Moon will be explored. "The symbolism of the Moon rendered fertile by the Sun," says Mircea Eliade, is related to "the regeneration of the *Nous* and psyche, the first integration of the human personality."[9] If we pursue the analogy further the Masculine, then, must be in service to the Feminine too, for the exact same reasons. And here, I am thinking of the Grail Quest. At this point, however, we meet Neumann again, for the Masculine is terrified of this power of the Feminine and reacts like the Left does with the Right – by usurping power, and, unable to see the larger picture, thinking that it can get away with it. Marduk may have been the first usurper but the process gained considerable traction with the Copernican Revolution. The emergence of pathological doubt is indeed the symptom of a coup staged by the left brain. For the left brain *cannot* believe, remember, all it can do is *know*. But it is a very particular kind of knowing. It knows by deconstructing, taking apart, separating, classifying, analysing, generalising in abstract and impersonal formulas. Mathematics is its language of choice. There is nothing wrong with this particular form of knowing, in fact it is an absolutely necessary piece of the jigsaw. It took us to the Moon and back. "It was a startling experience for me," says Joseph Campbell in *The Inner Reaches of Outer Space*, when "immediately before... Armstrong's landing on the Moon, ... Ground Control in Houston asked, 'Who's navigating now?' and the answer that came back was, 'Newton.'"[10]

But it is not the whole jigsaw, and it is potentially liable to a major pitfall. It derives general laws from particulars and once the laws are accepted, they are notoriously resistant to change, even in the presence of new particulars that challenge them. The Ptolemaic model is a classic example. It held for 1,500 years even though the motions of the retrograde planets were becoming increasingly problematic within the model. They were dealt with through the addition of more general laws – epicycles within epicycles – until the model could not hold any longer. Managing to escape fruitless left-brain loops, Copernicus only resolved the conundrum by a superb act of creativity outside the box – a typical right brain *modus operandi*. In most cases, however, the particulars that do not fit the model are discarded as irrelevant. Science prides itself on its ability to accept paradigm changes. Before it gets there, though, one has to contend with the left brain's absolutist stance and inability to eat humble pie. Then scientific inquiry crystallizes into dogma. Indeed, things go seriously awry when, wanting to impose its view – a view that is absolutely incapable of seeing beyond its limited scope – the left brain becomes dismissive of anything that contradicts it: "one would expect a dismissive attitude to anything outside of its limited focus, because the right hemisphere's take on the whole picture would simply not be available to it."[11] And lo and behold! A classic strategy used by hard-core rationalists is the sneer, the mockery, the dismissal with barely concealed contempt for anything that does not fit neatly into the straitjacket of their thinking. Everyone who has ever ventured outside the confines of pure rationalism knows what I am talking about. A sort of 'poor you', condescending attitude, mixed with scorn and derision. There is a price to pay for transgressing the explicit code of conduct of the left brain: it is to accept vulnerability and "to remain open, even to ridicule".[12] For as McGilchrist further emphasizes, "the left hemisphere pays attention to the virtual world that it has created, which is self-consistent, but self-contained, ultimately disconnected from the Other, making it powerful, but only able to operate on, and to know, itself."[13]

I have the utmost respect for Richard Dawkins. I am sure that his penetrating intellect has contributed much to the scientific debate, although, being no scientist myself, I am in no way qualified to judge it. He is also a very brilliant writer. There is, however, another side to Dawkins that is much valuable for the particular debate I am engaged in: it would be difficult, in my view, to find a more perfect embodiment of left brain thinking. I read *The God Delusion* because I wanted to understand where Dawkins was coming from – and, truly, it is most enlightening and often very funny. Dawkins has a great sense of humour, although it often is of the ironic ilk. Poking fun at – when he talks of "fairyologists", for instance, who are experts on "the exact shape and colour of fairy wings".[14] Funny, but a bit of a cheap shot really. It is, moreover, typical left brain sneering, an undermining, patronizing attitude towards people who are oh! so deluded and gullible. I wonder if Dawkins has read Kripal – he might find there a rather enlightening take on the hermeneutics of fairies. Scientific materialists, says McGilchrist, "share a sense of superiority, born of the conviction that others are taken in by illusion, to which those in the know have the explanation."[15] I have never seen fairies. I do not know if they exist or not, and probably asking the question in those terms is doing the matter a great disservice. I do believe, however, that a great many people have seen them, and I wonder what it means for us, humans. Richard Dawkins knows the answer – it is spelt out on page 34 of *The God Delusion*: "human thoughts and emotions *emerge* from exceedingly complex interconnections of physical entities within the brain."[16] It is all in the brain, the result of complex chemicophysical reactions. Matter somehow produces consciousness, feelings, ideas – and delusions. Fine, maybe – but is this more than a hypothesis? Has it been proven, without a shadow of a doubt, with due scientific rigueur? "Atheists do not have faith,"[17] says Dawkins. Well, they do. It is the faith of the left brain, a kind of stubborn, dogged, self-righteous belief:

A sort of stuffing of the ears with sealing wax appears to be part of the normal left-hemisphere mode. It does not want to hear what it takes to be the siren songs of the right hemisphere... It is as though, blindly, the left hemisphere pushes on, always along the same track. Evidence of failure does not mean that we are going in the wrong direction, only that we have not gone far enough in the direction we are already headed.[18]

"The existence of God," says Dawkins, "is a scientific hypothesis like any other."[19] And by God, Dawkins means god, goddess or any blend of divinities in no particular order. Should the existence of God be tested in the lab then? Well, no, absolutely not. It is the same misguided attempt as trying to test astrology in a scientific way. It is bound to fail because not everything is reducible to left brain processes, far from it. This attitude is called scientism – the unshakable conviction that the only valid epistemology is the scientific experiment. This conviction leads further and further into the 'prison of the mind', in Arendt's words:

The world of the experiment seems always capable of becoming a man-made reality, and this, while it may increase man's power of making and acting, even of creating a world, far beyond what any previous age dared to imagine in dreams and fantasy, unfortunately puts man back once more – and now even more forcefully – into the prison of his own mind, into the limitations of patterns he created himself.[20]

McGilchrist would say that we are in the presence of a typical left brain fallacy, absolutely unable to apprehend anything outside its own limited construct – a self-fulfilling prophecy, a classic left brain loop: God cannot be proven scientifically, therefore God does not exist and whoever disagrees is a moron of the highest order. But Science, really, is in no way equipped to tackle the problem of the existence of God, because Science and God are of a different order of knowledge:

> Many important aspects of experience, those that the right hemisphere is particularly well equipped to deal with – our passions..., all metaphoric and symbolic understanding..., all religious sense, all imaginative and intuitive processes – are denatured by becoming the object of focused attention, which renders them explicit, therefore mechanical, lifeless.[21]

Despite quantum physics, despite Heisenberg's recognition that the more man probed the universe, the more he encountered himself, what is cruelly lacking in the question is Jung's 'secret mutual connivance', Strieber's 'wink from the dark', a recognition of our collusion with the material, as well as a good dose of humility: our consciousness may be a reducing filter only able to apprehend what is within its scope. And we may have to admit that not everything is. In saying that, I am not advocating taking everything at *faith* value – God, Goddess, Witchcraft, whatever. Rather it is about keeping an open mind, a right hemisphere attitude that honours ambivalence.

Literalism might be the greatest flaw of 'left-brainism'. An inability to engage with the symbol and to think metaphorically. But one may ask – could literalism actually be the symptom of a paradox at the heart of left brain thinking? The left hemisphere, which cannot believe, deconstructs by doubting. But doubt is something that it is in fact quite uncomfortable with. Doubt is a strategy to arrive at black or white answers. Literalism might therefore be an escape route, the only way to reconcile the needs to simultaneously doubt *and* know. When the answers are not forthcoming, the doubting indeed becomes unbearable. The left hemisphere is then left dangling over a bottomless pit, and this more than anything is terrifying and has to be avoided at all costs.

Hence the vicious attacks, the self-righteousness, the dogmatic attitudes against everything that smacks of lunar ambivalence: "Astrology is neither harmless nor fun," said Dawkins in a famous 1995 *Independent* column, "and we should see it as an enemy of truth."[22] Keats talked of 'negative capability', the ability to "being in uncertainties, mysteries, doubt, without any irritable reaching after fact and reason" – a capability that is totally anathema to the left hemisphere. "Was Venus just another name for Aphrodite, or were they two distinct goddesses of love? Who cares?" asks Dawkins. "Life is too short," he adds, "to bother with the distinction between one figment of the imagination and many."[23] But you do not get rid of the Imagination unscathed.

Dawkins relates a childhood episode when he became frightened at night by what he thought was a ghost in his room mumbling abstruse words, only to realize that it was the wind playing tricks. "Had I been a more impressionable child, it is possible that I would have 'heard' not just unintelligible speech but particular words and even sentences. And had I been impressionable and religiously brought up, I wonder what words the wind might have spoken."[24] And with this statement he brushes aside religious experiences as the mere delusions of impressionable minds. But while I do not doubt that the wind was indeed the culprit, the physical phenomena behind Dawkins' fright, the question remains – why be scared in the first place? What does it say, symbolically, for the young Dawkins? To dismiss the experience, as Dawkins does, because "I had been told stories of priest holes in ancient houses, and I was a little frightened"[25] is not doing justice to the experience. In fact, if one stops for little more than one second, one cannot but be thoroughly puzzled by the reference to 'priest holes'. At the time of reading I did not know what priest holes were. Upon checking it turned out that priest holes were hiding places for Catholic priests persecuted during the Reformation. Now, here is an interesting motif! What is it about priest holes that had such an effect on the young Dawkins? How not to intuitively see that something very mysterious is at play? Dawkins is writing *The God Delusion* and feels the need to mention a frightening childhood experience involving persecuted Catholic priests? Could *The God Delusion* be the adult answer to the child's fear of priest holes? "Hush hush, there is nothing to fear… God is not hiding in the closet"?

What I am taking issue with is the typical left-brain denial of the symbolical. "We need metaphor or *mythos* in order to understand the world,"[26] says McGilchrist. But mythos is precisely what a mind in the grips of the left hemisphere is unable to apprehend. I remember once sitting in a café in London. At the time I was wondering whether I should study astrology or not – seriously study, I mean. I was debating with all sorts of pros and cons, not even sure that astrology was a serious subject in the first place. At the time, I knew close to nothing about it and no doubt my left hemisphere was desperately trying to infuse a dose of common sense into me. This café was a regular place for me to hang out. It did not have anything remarkable, except for good coffee, and the fact that on one wall hung what must have been thirty-five or forty pictures – drawings, cartoons, photos, a random kaleidoscope of colours and shapes, a kind of visual bric-a-brac. I was lost in thought when I mechanically raised my gaze to the wall. Immediately, my eyes fell on a picture that I had never noticed before. It was a simple drawing of the universe, dark with stars, and overriding it were the words 'THE FUTURE IS IN THE STARS'. At first, I did not pay much attention to it, only noting the funny coincidence of seeing this particular picture just when I was thinking about astrology. But then it hit me. There was my answer, clearly spelt out. The future is in the stars: not the fated future of Ficino's 'petty ogres', but *my* future – namely, the study of astrology. The writing was on the wall, literally and symbolically, although the left hemisphere, left to itself, would no doubt have dismissed it as nonsense. But I felt that I had received a genuine answer. Maybe it is at this time that I crossed over into the symbol, that is into an openness to recognise the little pebbles that one finds on the path. At the risk of delusion too, of apophenia – reading too much into the event – of wanting to believe too much, of being too eager, of twisting the arm of whatever is out there, or in there – us maybe – willing to provide guidance if we are humble enough to receive it, like an Apollonian 'Know thyself' Delphic injunction:

> The task is neither one of outwitting the stars, nor of simply suffering their 'accidental' intervention. Not atonement, but attunement. Know thyself becomes 'know the stars!'. Astrology as a way of living one's life in accord with the heavens.[27]

Astrology is not a superstitious tool to predict the future; it is an ethical practice to co-create it. It is about living a symbolic life, thereby striving to a profound alignment with the longing of the soul.

Literalism was understood by philosopher Owen Barfield as idolatry, a way to engage with the phenomena as objects independent of human consciousness, away from Original Participation: "The essence of original participation is that there stands behind the phenomena, *and on the other side of them from me*, a represented which is of the same nature as me."[28] This, Levy-Bruhl famously qualified as *participation mystique*. For Barfield, the story of modernity and science is a story of increasing objectivity, until we stood totally separate from what we were faced with, denying it any meaningfulness in its own right:

> The earlier awareness involved experiencing the phenomena as representations; the latter preoccupation involves experiencing them, non-representationally, as objects in their

own right, existing independently of human consciousness. This latter experience, in its extreme form, I have called *idolatry*.[29]

And so, said Hegel, "the oracles have stopped to speak, and the statues have become stone corpses."[30] And the Moon is a dead lump of rocks. "The happy bird-watcher," says Barfield, "does not say: 'Let's go and see what we can learn about ourselves from nature'. He says: 'Let's go and see what nature is doing, bless her!'"[31] Barfield called this stage Separation. We went to the Moon with this objectifying consciousness.

While this retreat from participation has greatly expanded an objective knowledge of the world, it has come with a heavy price. Caught in the 'prisons of the mind', "the more man learned about the universe," said Hannah Arendt, "the less he could understand the intentions and purposes for which he should have been created."[32] In the early part of the 20th century, German sociologist Max Weber highlighted the effects of a secular, highly rational society: "Our age is characterized by rationalization and intellectualization, and above all, by the disenchantment of the world. Its resulting fate is that precisely the ultimate and most sublime values have withdrawn from public life."[33] For Weber, science was chiefly responsible for this state of affairs – as the weapon of choice of a bloated left hemisphere that sake knowledge and control:

The growing process of intellectualization and rationalization does *not* imply a growing understanding of the conditions under which we live. It means something quite different. It is the knowledge or conviction that if *only we wished* to understand them we *could* do so at any time. It means in principle, then, that we are not ruled by mysterious, unpredictable forces, but that, on the contrary, we can in principle *control everything by means of calculation*. That in turns means the disenchantment of the world.[34]

For scholar Patrick Curry, who has extensively written on the notion of enchantment,

> modernity and enchantment… are related, but they are related as immiscible antinomies. Like oil and water, they do not and cannot mix… To put it another way, modernity as such necessarily entails disenchantment, and conversely, enchantment is the experience of a condition/world that is radically non-modern.[35]

"The most single hallmark of enchantment is *wonder*,"[36] says Curry, a point, he adds, emphasized by none other than Tolkien. In *The God Delusion*, Dawkins relates the story of a boy who, overwhelmed by his experience of nature, interpreted it in religious terms and became a clergyman. And now, says Dawkins, "that boy could have been me under the stars, dazzled by Orion, Cassiopeia and Ursa Major, tearful with the unheard music of the Milky Way, heady with the night scents of frangipani and trumpet flowers in an African garden."[37] In this lyrical line, a vivid artistic description of an African night, we witness Dawkins' poetic sensitivity. Is he not expressing wonder? He then continues: "Why the same emotion should have led my chaplain in one direction and me in the other is not an easy question to answer."[38] Dawkins is raising a very good point here. For the sake of argument, I will venture one possible answer: what if Dawkins was in a previous life one of those persecuted priests, who suffered so much for his faith that he came back with the radically opposite stance, making it his life goal to blow up to smithereens any whiff of religion? Isn't reincarnation (if one is so inclined to believe) the symptom of a deep longing of the soul to go through the whole spectrum of human experiences so as to grow and evolve? And, for a child of the Enlightenment, the iron hammers of Reason and Science would be the obvious tools of the trade to carry out that mission with gusto. Am I being too literal, trying too hard to find logical answers (sic) to the mystery of being?

But I am digressing. Finally: "A quasi-mystical response to nature and the universe is common among scientist and rationalists. It has no connection with supernatural belief."[39] I agree with Dawkins that many people of a strong rationalist bent display a deep reverence for nature without needing supernatural explanations. I have no doubt for instance that Brian Cox, the physicist and BBC journalist who has a particular dislike of astrologers and is not shy to express it, is in total awe of the sheer beauty of the cosmos. I can see it in his eyes.[40] As Owen Barfield is keen to emphasize: "The possibility of a selfless and attentive love for birds, animals, flowers, clouds, rocks, water permeates the whole modern mind, its science, its art, its poetry, and its daily life."[41] So, is modernity the enemy of wonder? I am not sure about that. But if not through a loss of wonder, how is the world disenchanted then?

For McGilchrist, in a world gripped by the left hemisphere "it would become hard to discern value or meaning in life at all; a sense of nausea and boredom before life would be likely to lead to a craving for novelty and stimulation."[42] Loss of meaning... This might indeed be the price we paid for modernity, whose main thrust is based on the notion of 'progress'. This was recognised by Weber a hundred years ago, who wondered aloud whether progress did not carry in itself the seeds of meaninglessness, in particular in our relation to death:

A civilized man... who is inserted into a never-ending process by which civilization is enriched with ideas, knowledge, and problems may become 'tired of life', but not fulfilled by it. For he can seize hold of only the minutest portion of the new ideas that the life of the kind continually produces, and what remains in his grasp is always merely provisional, never definitive. For this reason death is a meaningless event for him. And because death is meaningless, so, too, is civilized life, since its senseless 'progressivity' condemns death to meaninglessness.[43]

In the age of Internet and virtual reality, these words sound more prophetic than ever. This arc of progress, the straight line of development towards better and stronger, the gnawing dissatisfaction with life desperately dulled by a craving for novelty, the fascination with technological tools, the meaninglessness of death, are typical of the solar consciousness that we have experienced for the last three hundred years or so. "There persists..." said Alfred North Whitehead,

> the fixed scientific cosmology which presupposes the ultimate fact of an irreducible brute matter, or material, spread throughout space in a flux of configurations. In itself such a material is senseless, valueless, purposeless. It just does what it does do, following a fixed routine imposed by external relations which do not spring from the nature of its being.[44]

Written in 1925, just as quantum physics was bursting on to the scene, these words are still relevant today. It seems that the lessons of the *Uncertainty Principle* are just too challenging to be accepted by culture. But this cosmology has taken us to the Moon too, and in this epochal adventure we saw the capacity of the modern mind to stand in awe and even be enchanted. Maybe because, as Curry says, "in the interstices of the grid, among places and people... overlooked by power, and even, I daresay, in the hearts of many of its servants, enchantment still lives."[45] But has the Moon landing given meaning to our lives? Or has it actually contributed to a deeper sense of alienation and disenchantment? As scholar Tom Cheetham remarks: "the defining characteristics of modernism is its loss of a cosmological sense of wholeness, of a hierarchical continuum that includes the verticality of transcendence."[46]

I want to argue that sticking to a pure mechanistic and scientific reading of the *Apollo* mission is not meaningful, in the sense that it does not provide us with a deeper sense of meaning. Maybe what is needed is, as Tom Wolfe puts it, a "philosophy of space exploration". Frank White, who coined the term 'Overview Effect',[47] considers that "we are not simply reaching out into space to use extraterrestrial resources and create opportunities here on Earth. Rather, we are laying the foundation for a series of new civilizations that are the next logical steps in the evolution of human society."[48] In so doing, he adds, we may be "performing a vital function for the universe as a whole."[49] But while this impulse to colonize other worlds may satisfy the explorer in us, does it assuage our existential angst? Does it answer the question, "What does it mean to be human?" For White, seeing the Earth from a distance in space will have the effect of altering consciousness to such an extent that future colonies of humans on alien worlds will be radically transformed, with a profound sense of unity at their core underpinned by the Overview Effect. Maybe. That is putting a lot of faith in the Overview Effect.[50] Before fantasizing about other planets and seeing pies in the sky, however, maybe we could start by altering our relationship with our own, profoundly damaged, Earth.

"So what?" said Yale students to Joseph Campbell after the Moon landing – which infuriated him so much. The 'So What' attitude may be what best defines our world, hiding as it does behind a "self-protective carapace of ironic knowingness and cynicism",[51] typical, says McGilchrist, of the left hemisphere. Incidentally 'So What' is also the name of the opening track on Miles Davis' 1959 seminal album *Kind of Blue*. I read once an interview by Dennis Hopper who claimed ownership of the title. As a young man he was hanging out a lot with the New York jazz music scene, said he, and, as 'so what' was his signature phrase, it was used by Miles Davis. Well I don't know, I suspect Dennis Hopper to be a bit of a trickster, so am not sure what to make of it. I like the story though.

Meaning, essentially, needs context, and this, says McGilchrist, is the domain of the right hemisphere:

> For the same reasons that the right hemisphere sees things as a whole, before they have been digested into parts, it also sees each thing in its context, as standing in a qualifying relationship with all that surrounds it, rather than taking it as a single isolated entity. Anything that requires indirect interpretation, which is not explicit or literal, that in other words requires contextual understanding, depends on the right frontal lobe for its meaning to be conveyed or received.[52]

Apollo needs his brother Hermes, who made a lyre out of a turtle shell and gave it back to Apollo, for him to *enchant* us. "The orientation of the meaning of the whole has been lost to such an extent," said Neumann, "that today it is already almost forbidden to pose the problem of meaning at all."[53] We too, then, need to see a lyre in the shell of the *Apollo* mission. Only then can this whole epic journey reveal its meaning to us and re-enchant our lives.

The hermeneutic loop, again, the never-ending spiralling of consciousness – a stairway to the stars, twisting its way among fallen angels and empty whisky bottles.

Chapter 5

The Argonauts

The Lessons of the Argonautica for our Journey to the Moon

The whole process, the whole Argonaut journey, had failed, essentially because of the disregard of the feminine element.
– E. Edinger

The word 'astronaut' comes from the Greek 'astro' – star – and 'nautes' – sailor, itself originating from 'nau', boat. An astronaut is therefore a star sailor. The word was first used in its modern sense in 1930 by writer Neil Jones in a short story entitled *The Death's Head Meteor*. The etymology of the word 'astronaut' leads back to the Argonauts – the sailors of the vessel *Argo* who went in search of the Golden Fleece. The mythical status of the Argonauts was not lost on NASA:

> On 9 April 1959 NASA announced selection of the seven men chosen to be the first US space travellers, 'astronauts'. The term followed the semantic tradition begun with 'Argonauts', the legendary Greeks who travelled far and wide in search of the Golden Fleece.[1]

The quest for the Golden Fleece was said to be a favourite of the alchemists, who saw in it an allegory of their own quest for alchemical Gold and who held Jason as one of their own. I want to explore this story in some detail because I see in it a very meaningful parallel with our modern tale of the astronauts going into space. The old myth may highlight something important with regards to our own mythical flight into space. For this task I am using two principal sources. While, as usual, several versions of the Greek myth are available, I am indebted to the *Argonautica*, written by Apollonius of Rhodes in the 3rd century BCE. And in parallel I am drawing on Edinger's psychological interpretation of Greek myths developed in 1994 in *The Eternal Drama*.

The basic storyline of the *Argonautica* is classically Greek. Trying to rightfully recover his throne, the solar hero Jason is given an impossible task by the usurper Pelias: he must recover the Golden Fleece, a mythical ram's hide hanging in a faraway land, a land that Jason in all likelihood will never reach alive. To accomplish this feat Jason gathers all the heroes that Greece can conjure up, amongst whom stand out Herakles, Orpheus, and the Dioskouroi, Polydeukes and Kastor (a.k.a. Pollux and Castor, the Gemini brothers). A sort of supergroup, an Ancient Greek ancestor to the *Traveling Wilburys*.[2] Even the gods are impressed by this display of male power: "all the gods looked from heaven upon the ship and the generation of demi-gods who sailed the sea, best of all men."[3]

In its teething years, NASA seemed to have had a thing for mythology: the first three US space programmes were respectively called Mercury (1959-1963), Gemini (1961-1966) and Apollo (1960-1972). Each programme had a specific objective to accomplish. The task of 'Mercury' was to conduct the first US manned space missions – the first visit of man into space. Quite sensibly, NASA's Abe Silverstein saw a parallel with the God of travellers, with his legendary winged sandals and his ability to navigate the worlds – the only God who could visit Olympus as well as Tartarus. Mercury it was, then. The first astronauts were nicknamed the Mercury Seven and very rapidly, the programme became associated with the number seven:

In each of Project Mercury's manned space flights, the assigned astronaut chose a call sign for his spacecraft just before his mission. The choice of "Freedom 7" by Alan B.

Shepard, Jr., established the tradition of the numeral "7," which came to be associated with the team of seven Mercury astronauts. When Shepard chose "Freedom 7," the numeral seemed significant to him because it appeared that "capsule No. 7 on booster No. 7" should be the first combination of a series of at least seven flights to put Americans into space.[4]

For esotericist John Michell the number '7', as a quarter of the Moon cycle, has an indirect association with the Moon:

> Seven is unique among the numbers of the decad because, as the Pythagoreans said, "it neither generates nor is generated", meaning that it cannot be multiplied to produce another number within the first ten, nor is it the product of other numbers. For that reason the heptad was called the Virgin and was a symbol of eternal rather than created things. It was particularly related to the measurement of time, the seven ages of man, and the seven days of the week. It is thus connected with the perfect number 28, which governs the periods of the moon and the female.[5]

Is there an esoteric subtext behind the '7' motif of the Mercury programme? Outwardly, the Mercury Seven were quite willing to portray themselves as deadpan rationalists. When first introduced to the press, the astronauts were asked about their religious faith and how it was likely to support them in the challenges ahead. "I think I should like to dwell more on the faith in what we have called the machine age,"[6] declared astronaut Wally Schirra. Was it a smokescreen to hide more esoteric leanings?[7]

NASA upped the ante with the Gemini programme, which was based on a two-man spacecraft. Its main objective was to develop the technique of rendezvous in space with another spacecraft – a technique that would prove vital for the Moon landing and which may have greatly helped Buzz Aldrin, who wrote his PhD thesis on orbital rendezvous, hitch a ride aboard Apollo 11. The name 'Gemini' was chosen because

'Gemini', meaning 'twins' in Latin, was the name of the third constellation of the zodiac, made up of the twin stars Castor and Pollux. To [NASA] it seemed an appropriate connotation for the two-man crew, a rendezvous mission, and the project's relationship to Mercury. Another connotation of the mythological twins was that they were considered to be the patron gods of voyagers.[8]

You have to take your hat off to NASA. They got their mythology pretty spot on. It is almost mystifying in fact. The reference to the third constellation of the zodiac, the oblique acknowledgement of Mercury ruler of Gemini – what has this astrological jargon got to do with scientific space exploration? If you wanted to split hairs, you might say that NASA had jumped the gun a little in referring to the twins as the "patron gods of voyagers", since they were known, rather, as the gods of horsemanship and *protectors* of travellers, but that would be just pedantic. But why this insistence on mythological figures? *Why mythologize the missions?* Is this the symptom that somewhere, in the recesses of Jung's collective unconscious, mythology is still alive and kicking, as relevant today as it was two thousand years ago? Or is it just mere fun, a nod to a distant past lost in the fogs of history, an innocuous wink? There is no doubt, however, that by plugging into the archetypes NASA was duly inviting the Gods on board – that is, reaching into a deep and ancient human experience. In so doing NASA was acknowledging the mythical dimension of the space programme, way beyond technology and politics. There was the recognition of a human adventure that, although obviously quintessentially American, was encompassing the whole of humankind.[9] This became especially true after President

Kennedy firmly set the eyes of the nation on the Moon. With *Apollo*, NASA became invested with a kind of divine mandate, embarking on an adventure that would redefine the place of man in the universe. Armstrong strikingly encapsulated the nature of this endeavour with his first few words on the Moon. After *Apollo*, NASA did not send men into space until the shuttle was up and running. Henceforth the different NASA missions stopped carrying the same mythical charge,[10] becoming much more neutral – Skylab, Viking, Pioneer. Even the shuttles' names – Discovery, Challenger, Columbia – dropped the mythical connection. It is as if, after the successful Moon landing, it was NASA itself that acquired a mythical status. After *Apollo*, there was no need to tap into myth any longer: NASA *was* the myth. The astronauts, too, had acquired a mythical status. As Frank White put it in the *Overview Effect*: "The astronauts fit into the mythical subconscious archetypes of the gods and heroes of old, flying beings who perform feats of daring no one else is able or willing to do."[11]

But a remarkable development is now taking place: with the upcoming *Artemis* missions, NASA is reverting back to mythology. The eyes are now set on Mars, with the Moon a mere stopover on the way. By venturing further away than ever before, the *Artemis* missions will set mankind on a new course. Fifty years on, humans will leave the orbit of the Earth once more and NASA taps into myth again, as if the outer exploration of new space frontiers was inevitably calling forth a deep and ancient archetypal experience. As above, so below: the higher we go, the deeper we reach. A magical act it is too. By invoking Artemis, NASA summons by association her twin brother Apollo, replete with his sparkling laurels: the most successful space mission ever designed.

Which begs the question: is NASA much more esoterically conversant than is acknowledged? Was there a conscious invocation of the Gods with the *Apollo* missions? But if there ever was, esotericists and magicians know full well that you do not invoke an archetype lightly, for the Gods do not work for free. I cannot help but be struck by the fact that the very first *Apollo* mission – *Apollo 1*, which was a ground mission to test the capsule – ended in such dramatic archetypal fashion. The God of the Sun got his sacrificial victims: the three astronauts locked inside the cabin tragically burned to death in a devastating fire that shook the programme to the core. Apollo is a vengeful God and will seek retribution if the offerings are found wanting. The higher echelons of NASA were seemingly ripe with masons and nothing stirs the conspiracy pot more than freemasonry. Was something more magical and esoteric at play in the space programme? Rather than his rendezvous credentials, was being a mason what earned Buzz a place on *Apollo 11*? "The practice of magic," said Jung in *Liber Novus*, "is to make what is not understood understandable in an incomprehensible manner."[12] That sounds like a pretty sensible definition to me. Conspiracy theories are equally baffling and fascinating. They can suck you down the rabbit hole at breakneck speed but they, too, have a story to tell. They do wink back from the dark, showing the human mind at play. "If the doors of misperception were cleansed, one would see everything as it is – infinitely opaque!", they whisper, parodying Blake. But they may also be no more than the typical paranoia of the left brain, always wary of being lied to, deeply entrenched in the certainty that it knows better and cannot be deluded.

Whatever NASA's esoteric proficiency, one cannot fail to see a strong parallel between the Astronauts and the Argonauts, who both embody a rather classic solar consciousness: virility, adventure, danger, focus, excellence, professionalism, as well as a rugged male camaraderie. Think *The Right Stuff*. In this regard, *First Man*, Neil Armstrong's 2018 biopic, offered an interesting shift in perspective. By shedding light on Neil's wife Janet, it openly wondered what it meant to live as an astronaut's spouse, with the constant shadow of death lurking in the corner – and God(dess) knows that Neil had his share of close encounters with the Grim Reaper! Apollonius of Rhodes did not show the same consideration for the Argonauts' female companions, who were totally absent from the storyboard. This may be as much as we have come in the last two and a half thousand years: while men retain their heroic status, we now wonder aloud what it is like to be a hero's WAG – although Homer, ever the trailblazer, did make room for Penelope's angst.

Let's now dispense with the suspense: just like Armstrong on the Moon, Jason did succeed and brought back the Golden Fleece. Unmitigated triumph? Another tale of a solar character accomplishing a heroic task, complete with the compulsory fight with the dragon? Outwardly yes, but in truth anything but. A great lesson has been ignored, and a great crime perpetrated, by Jason: the wronging of the Feminine. Despite his pledge to the sorceress Medea, without whom he would never have succeeded in his task, Jason eventually abandons her to marry the princess of Thebes – just like Theseus abandoned Ariadne on Naxos after defeating the Minotaur. "The Masculine," says Edinger, "could not achieve a balance with the Feminine, hence the most that could be done was to exploit the Feminine principle and then drop it again. The result was that the anima turned into bitterness."[13] Discarded, spiteful, Medea becomes a raging fury, killing the princess of Thebes and turning on her own children, finally vanishing into the sky in a chariot pulled by a dragon. In so doing she joins Ereshkigal and Lilith in the pantheon of exiled female deities: the dark feminine characters unacceptable to solar consciousness, who will yet have to be somehow reintegrated if we are to honour the psyche in her wholeness.[14]

Much like Armstrong, but for different reasons,[15] Jason is a rather unusual solar hero. At times seemingly weak, indecisive and ineffectual, he cuts a forlorn figure when compared to the likes of Achilles, Hector or Perseus. His very legitimacy as a leader is questionable. When the heroes debated before departure about leadership, it was not to Jason that they turned but to the greatest fighter of them all, Herakles. "Let no one offer me this honour – I shall not accept it," said Herakles, "let he who gathered our band together lead us on our way."[16] Was Jason a leader by default then?

It reminds me of a story from my student years. At university, it was the custom to elect every year a Board of Students to take charge of campus life – essentially organising parties. Competing groups of students were vying for the honour. I was in one such group, and it turned out that we were the only ones applying for the job – so there was no real doubt as to the result of this Soviet-like election. The problem, however, was internal to the group – two students were competing for leadership, two Herakles figures going head to head. In contrast to the Herakles of the *Argonautica*, though, none of them was in any way willing to relinquish his claim to the throne. One famous evening, we gathered to elect our *Lider Maximo*. To compound the problem, our group was composed of 10 members. We voted, re-voted, and re-re-voted. The score was desperately tied 5-5. No one would budge. This little tit-for-tat game went on for several hours. At 3 or 4am, with everyone thoroughly exhausted, a radically different proposal was suddenly suggested to break the deadlock. The grand plan was to elect *me* as Leader – but as a straw man, a consensual but ineffectual figure, while the two contenders would remain in the background steering the boat. It was a dreadful idea. I had no desire to be the leader of anything, even less so a puppet leader, and it was bound to fail, for the drama at the heart of the problem had not been addressed. But hey – everyone thought it was a brilliant idea, and I joined the chorus. I accepted. What exhaustion and tiredness can do to you! At some level I must have been flattered too, although in truth the offer could have been construed as pretty insulting to me. Anyway, the farce lasted for ten days before I came to my senses and removed my puppet mask. It is lying in a corner of my psyche, all dusty and wrinkled. For the remainder of my student years, I became known as the 'straw man'. Meanwhile, in a cathartic moment worthy of the best Greek tragedy, one of the two contenders plainly said that if we did not choose him, he was leaving, plain and simple. He was then duly elected because, in truth, he was a better choice. When push came to shove it was the obvious decision. All along I voted for the other candidate out of sheer loyalty, even though I felt in my heart that he was not the best option. After this episode, I knew politics were not for me.

But kudos to Jason. He is not your typical ambitious and arrogant solar character. During the eventful trip to Colchis he does indeed display a lot of temperance and gentleness, a capacity to listen and to forgive, even when he is seriously challenged by his men: "Son of Aison [a.k.a. Jason], has Fear come over you and crushed you with its weight?" asks the blasphemous Idas spitefully, "It is this which panics men who are cowards."[17] I do not know of many males who would graciously take the insult of cowardice. Even less so male leaders. But Jason does not react and it is never entirely clear in the story whether it is out of weakness or out of true inner strength. His final behaviour towards Medea of course does not precisely endear him to us. As much as it pains me, I have to admit that I have some affinity with Jason's flawed character. I, too, have wronged the Feminine, although I am hesitant to venture further into this tricky territory. This may be a story for another time. Does this kinship with Jason, this partnership in crime, make me fit to talk about the wronging of the Feminine – or particularly unfit, rather?

Scholar Richard Hunter has a kinder take. While acknowledging that Jason was driven by events rather than driving them, he sees in him a Ulysses-kind of hero, relying on *metis* – wisdom, cunning – rather than on brute force, a hero ready to use any trick in the toolbox to come back home. Hunter also insists on the super nature of many of Jason's companions, with whom Jason necessarily has to share the limelight, and on the communal aspect of the *Argonautica*. As for leadership, however powerful Herakles may have been, says Hunter, he was not cut out to be a leader of men, his path being a very individual one – and in fact, Herakles is early on in the story mistakenly left stranded onshore by the crew of the *Argo*, whose epic journey then unfolds without him. The best men do not always make the best leaders and in all likelihood a hot-blooded character like Herakles would have clubbed Idas to death for hurling his insult, with a great risk of jeopardizing the peace on board and literally rocking the boat. Jason is a consensual leader and may have a more feminine way of leading. Like Laura Roslin in *Battlestar Galactica*. The art of gentle persuasion – especially when surrounded by a bunch of strong, overbearing egos – is, arguably, more effective.

Despite winning the fleece, though, Jason eventually fails, and with him the whole venture. For Edinger, the script was written before the voyage even started, its demise foretold in the story of the Golden Fleece itself. For, indeed, what is this mysterious Golden Fleece? How did it come about and why did it end up hanging on a tree in modern-day Georgia?

The ball was set in motion in Greece's Boiotia way before Jason's birth. Phrixos and Helle were the offspring of King Athamas who, after their mother's death, remarries with Ino. Ino is not a nice stepmother – a motif strongly entrenched in our culture, right down to Snow White and to cheap mother-in-law jokes. To get rid of the children, Ino tricks Athamas into believing that he has to sacrifice Phrixos and Helle to Zeus to save his land from famine. One is often confounded by the ease with which men are manipulated by women in myths and fairy tales and I wonder. Is this a patriarchal take on the deceiving ways of Woman, or a matriarchal view of the ineffectiveness of Man? Anyway, Zeus and Hermes come to the rescue of the siblings before they are sacrificed on the altar. A magical, golden-fleeced ram scoops them up on its back and flies away. Shortly after take-off, however, Helle slips off the back of the ram and falls into what henceforth became known as the Hellespont – the sea of Helle. Phrixos, however, completes the journey and lands on the far side of the Black Sea, in Colchis. Once there, his first move is to thank the ram by promptly sacrificing it to Ares, with the Golden Fleece hung on a tree as a shrine. Phrixos ends up marrying Chalkiope, Medea's sister, and the loop back to Jason is closed. For Edinger, the fall of Helle was a harbinger of what was to follow:

The ram with the golden fleece signifies a masculine aspect of the Self. Its golden fleece suggests its supreme value, but its masculine character indicates that it is only a partial expression of the Self, and that limitation runs through the entire myth: much is accomplished, but incompletely. We recognize this theme from the very beginning in the one-sided nature of the symbol and in the early loss of the girl child Helle. The feminine element is lost repeatedly and then at the end, in retaliation, destroys the whole enterprise.[18]

The wronging of the Feminine is a theme that echoes throughout the *Argonautica*. Very early on in the journey, the Argonauts land on the shores of Lemnos. There, they encounter an island exclusively populated with women. Apollonius curtly explains why: "in the year just passed, the whole people had been pitilessly killed in one stroke by the wickedness of the women."[19] What happened? Why were the women pitiless and wicked? We learn that after raiding the Thracian mainland, the men of Lemnos took a fancy to slave-girls, deserting their wives and their fatherly duties. "O wretched women whose grim jealousy knew no bounds!"[20] says Apollonius. Well, that is harsh! The women, of course, may be said to have overreacted a tad: they

117

"destroyed not only their husbands together with the slave-girls in their beds, but also the entire male population so that there could be no requital in the future for the awful murder."[21] The Feminine obliterated the Masculine but in so doing jeopardized its very own survival, for the women are now barren. When the *Argo* lands on their shores, therefore, the Lemnian women invite the Argonauts into their beds for the sake of the children they could father. The heroes, all of a noble character, politely accept the invitation but subsequently display the greatest resistance to be summoned back on to the *Argo* to pursue their journey. For Edinger, this episode symbolises a classic pitfall on the road to individuation: "When there is a first reconciliation of opposing factors, a strong temptation arises to succumb to the pleasant urge, to settle for that, and to forget about the goal that is still far distant."[22] We find an echo of this theme in the *Odyssey*, although this time it is not women but narcotic fruits that enchant the heroes. Early on in the journey Ulysses and his men land in the country of the Lotophagi, eaters of Lotos fruits whose artificial paradises prove irresistible. The men have then to be forcefully dragged on board by Ulysses:

> The trees around them all their fruits produce
> Lotos the name, and dulcet is the juice!
> (Thence call'd Lotophagi) which, whoso tastes,
> Insatiate riots in the sweet repasts
> Nor other home, nor other care intends,
> But quits his house, his country and his friends.[23]

The episode of the Lemnian women anticipates the Medea motif. The wronged Feminine, scorned and insulted, brings the castle down, and everyone in it. Needless to say that this feminine rage is today, more than ever, for everyone to see. Feminists burnt their bras in the 70s, in the digital age their anger is now expressed through the #metoo campaign – in what is probably a wider and deeper movement since it reaches beyond avowed feminists. And this is only the tip of the iceberg. One can sense a seething resentment boiling up in many modern women, outraged by four thousand years of patriarchy. Many men, too, are longing to reconnect with the Feminine – their inner Feminine especially – for they intuit that an excess of Sun has terribly impoverished them.

The relation between Jason and Medea is at the heart of the story. Without Medea, Jason would not have succeeded in obtaining the fleece. In true Jungian fashion, Edinger associates Medea with Jason's anima: "this help from the anima, who had contact with arcane powers – we could say, with deep layers of the unconscious – was indispensable to his success."[24] In the myth, Medea helps Jason not once, not twice, but repeatedly, and at each step the hero seems to take less and less responsibility – a fatal flaw leading to disaster. Jason's first task is to yoke together two fire-eating brazen bulls, plough a field with them, sow dragon's teeth into the field and fight the armed men springing from the sowing. A mighty endeavour by any standard – but Jason is given a magic ointment by Medea, making him invulnerable for one day. Furthermore, he is also given a tip by his crafty lover: throw a stone in the midst of the soldiers who will then fight among themselves for its possession, annihilating each other in the process. Jason goes into the fight "naked, and in different ways [resembling] both Ares and Apollo of the golden sword."[25] A proper solar hero going into battle, a commanding image that he will never again manage to emulate in the story. "What seems to be indicated by all this," says Edinger, "is that the ego must expose itself to the primordial powers of the masculine archetype represented by the fire-eating bulls, and the opposites that arise in the form of armed men must also be dealt with."[26] Victorious, Jason must then recover the Golden Fleece in the sacred grove. The Fleece is guarded by a fierce dragon: "directly in front of [Jason and Medea], the dragon stretched out its vast neck when its sharp eyes which never sleep spotted their approach, and its awful hissing resounded around the long reaches of the river-bank and the broad grove."[27] The dragon must be fought and vanquished, a very classic theme.

There are two ways to look at the archetypal fight with the Dragon: as murder, or as necessary sacrifice. As Cashford and Baring emphasize,

> when the hero myth is perceived in terms of the growth of consciousness, it becomes an inner quest for illumination. Here the conflict is not so much between good and evil, but rather one between a greater or lesser understanding... The aim of the hero is to master his own inner darkness – his fear, or the limitation of his knowledge.[28]

In this light, the killing of the Dragon is not a wanton act of destruction but a necessary stage in the growth of consciousness. This echoes the positions of Campbell and Neumann: "the development of the conscious system, having at its centre an ego [a hero] which breaks away from the despotic rule of the unconscious, is prefigured in the hero myth."[29] The difference between murder and sacrifice revolves around the notion of *transformation* – one of the most obvious teachings of the Moon. "Transformation," says psychologist Marion Woodman, "moves energy from the unconscious to consciousness... Where the slaying of the dragon is not understood as a symbolic process of transformation, then the feminine is separated from its own source of life and power in matter (*mater*)."[30] For Woodman, there is no doubt that something has gone seriously amiss in the hero's myth, as is vividly seen in our pathological materialism:

> The myth of the sun hero fighting the dragon and winning his way to consciousness has suffered from overkill. The energies of that myth have been exhausted and we are now struggling

with the abuses of its excess... The effort of centuries to kill the dragon has ended in the worship of mother in concrete materialism... When dragon slaying is concretized, mother becomes concrete matter.[31]

This life-denying attitude is seen in our destructive relation to Mother Earth and in this, Woodman joins the ranks of ecofeminism: "the individual relationship to the feminine has extended into a collective attitude to the planet... This disturbing situation is in large measure the result of a flawed solar myth that confers upon the masculine a heroic status, which now threatens us with extinction."[32]

James Hillman concurred with Woodman that it was high time we moved past the Hero's myth, which "has become for our psyche the myth of inflation and not the secret key to consciousness."[33] But Hillman was also turning the tables on Neumann by challenging the association of ego with consciousness:

The Hero-myth tells the tale of conquest and destruction, the tale of psychology's 'strong ego', its fire and sword, as well as the carrier of its civilization, but it tells little of the culture of its consciousness. Strange that we could still, in a psychology as subtle as Jung's, believe that this King-Hero, and his ego, is the equivalent of consciousness.[34]

For Hillman, an ego-consciousness "limits consciousness to the perspectives of the [Platonic] cave which today we would call the literalistic, personalistic, practicalistic, naturalistic, and humanistic fallacies."[35] Hillman wanted to de-locate the seat of consciousness to the soul and give back its due to the Imagination: "consciousness arising from soul derives from images and could be called imaginal."[36] Inevitably this led him to an encounter with the anima ('the soul', in Latin) but he did assign to the anima a much wider role than what he felt was Jung's polarized view – a view according to which the masculine ego opposed the feminine 'other', and the conscious met an unconscious personified as anima. Hillman saw this tension played out in the fight with the Dragon, with the anima its chief victim:

> The entire relationship with anima is placed into the mythologem of the heroic ego and his archetypal fight with the dragon. Then efforts to integrate, 'to bring these contents to light', become a depotentiating of personifications and of their imaginal power, a drying-up of the waters, and a slaying of the angel (seen to be a dangerous fairy-demon by the ego).[37]

For Hillman, then, the slaying of the dragon – in reality an angel demonized by the ego – denies and prevents soul-making, with the result that the ego is kept in its heroic stance. A severing of the connection to the imaginal then results, with dramatic consequences. An excessive Sun dries up the moistness of the Moon. This is TS Eliot's *Waste Land*:

> Here is no water but only rock
> Rock and no water and the sandy road
> The road winding above among the mountains
> Which are mountains of rock without water.[38]

Jason has a very different kind of encounter with the dragon. It is Medea who in effect takes the initiative, while, says Apollonius, "behind her followed the son of Aison, terrified."[39] She puts the dragon to sleep with her magic: "with a fresh-cut sprig of juniper which had been dipped in a potion, Medea sprinkled powerful drugs over [the dragon's] eyes while she sang, and all around sleep was spread by the overwhelming scent of the drug."[40] Then, effortlessly, "Jason removed the golden fleece from the oak at the maiden's instructions."[41] By then, one has the distinct feeling that Jason has abdicated any attempt at anything resembling volition. By not confronting the dragon, however, the possibility of transformation is unavailable to him. From there on things go from bad to worse. When the Argonauts and Medea sail away on the *Argo*, desperately trying to escape the wrath of Medea's father, it is Medea, again, who comes to the rescue by sacrificing *her own brother*, Apsyrtus, dismembering him and throwing the body parts into the sea so as to slow down her father. "Although the image is repulsive," says Edinger, "it represents the dissolving of a certain concretization of libido so that it may become available for a new kind of relationship."[42] You have to love archetypal psychology. Medea performs another service for Jason once the pair returns to Greece. The usurper Pelias has killed Jason's parents: vengeance is due and Medea tricks Pelias' daughters into killing their own father. Jason, once more, blissfully stays out of the fray and has Medea do the dirty work for him. This cannot end well and it does not. Betrayed by her lover Medea disappears in a ball of fire and Jason, having lost the favour of the gods, ends up a homeless wanderer. The whole venture, indeed, has failed. The anima has become bitter and the ego, cut off from its life-giving source, is destitute.

The wronging of the Feminine is a theme that, quite logically, has been extensively covered by feminist authors. But feminism is a maze of a subject, with many strands and counter strands, and I want to circumscribe it to one of its many incarnations, ecofeminism. Ecofeminists see an obvious parallel between the submission of the Feminine and the exploitation of Nature. In 1980 Carolyn Merchant published *The Death of Nature*, a self-styled essay on *Women, Ecology and the Scientific Revolution*: "the... image of nature as a female to be controlled and dissected through experiment legitimated the exploitation of natural resources."[43] Equating women with nature is anathema to many feminists, who see in it no more than a patriarchal plot to justify the inferiority of the Feminine and thereby keep women in their place: "If women overtly identify with nature and both are devalued in Western culture, don't such efforts work against women's prospects for their own liberation?"[44] asks Merchant. Maybe the core issue is that we keep devaluing nature in the first place, and here Merchant lays the blame squarely at the feet of the Scientific Revolution and the scientific mind that underpins it. Up to the Renaissance, she adds, "an I-Thou relationship in which nature was considered to be a person-writ-large was sufficiently prevalent... that it could effectively function as a restraining ethic."[45]

The notion of a 'I-Thou' relationship was developed by Austrian philosopher Martin Buber in the 20th century. For Buber, humans could encounter the world in two major ways: as a Thou or as an It. "The attitude of man is twofold in accordance with the two basic words he can speak."[46] In the 'I-Thou' encounter, the 'I' does not stand separate as an isolated unit but needs the 'Thou' to become:

> The basic word I-Thou can be spoken only with one's whole being. The concentration and fusion into a whole being can never be accomplished by me, can never be accomplished without me. I require a Thou to become; becoming I, I say Thou. All actual life is encounter.[47]

Ecophilosopher Thomas Berry had a beautiful way of expressing a similar idea: "the universe is a communion of subjects, not a collection of objects". In South Africa we find the concept of *Ubuntu*, which can be expressed as "I am what I am because of what you are". 'I' need 'You'. Nelson Mandela and Desmond Tutu often referred to *Ubuntu* as a way of being in the world. The word *Ubuntu* has a particular resonance for me. Immediately after leaving my job as an oil trader I was put in touch with a group of people in the initial stages of launching Ubuntu Cola, the first Fairtrade Cola to be sold in the UK. Initially, I was feeling a bit uncomfortable with the use of the *Ubuntu* humanist philosophy as a marketing tool to sell a fizzy drink but the project was already well under way. Besides, one of the objectives of the company was to reconcile business and ethics, if such a thing is at all possible, and the Ubuntu name symbolised this commitment. I joined the budding entrepreneurs, and my first task was to fly to Malawi to meet the Fairtrade sugar producers. But this is another story. I want to tell the anecdote of how I made the improbable jump from Crude Oil to Fairtrade Cola – which, apart from their colour, do not share much in common. The seed was planted five years before – Christmas 2002. One lazy afternoon, I was browsing the *Independent* newspaper when my eye caught an article pitching a new project backed by Oxfam: Progreso was to be a Fairtrade coffee chain, a counterpoint to the then classic American or British coffee chains that did not have Fairtrade on their radar. I had always felt an affinity to the Fairtrade concept and, on impulse, I decided to contact the project manager at Progreso. I met him in a pub and after our discussion I decided there and then to support the project. He was quite surprised – who was this guy springing out of nowhere like a jack-in-the-box and parting with his money without imposing the usual due diligence that donors always insist on? A handshake and a cheque, that was it. I never saw this man again – nor was I planning to. Progreso developed a bit before sadly crashing down to earth. Fast forward five years. I had decided to leave the oil market and I needed to reinvent myself. I was thinking of Fairtrade as a potential avenue – after all, in 'Fairtrade' we find 'trade', a line of work I was familiar with, although I was not so sure about how to do 'fair'. I was pondering how to engineer my move when I suddenly remembered the Progreso man, who was still working for Oxfam. I caught him just in time. He was about to leave the UK to settle in Cambodia but his parting gift was to get me in touch with the Ubuntu people – and the rest is history.

Sometimes we plant seeds without knowing we are doing so – no agenda, no plans, and the fruit falls from the mature tree five years later. Life works in unexpected ways – occasionally magic does happen, catching us blissfully unaware. Maybe I am telling this story to remind myself that surprises can be good, too.

Merchant identifies the Scientific Revolution as the point at which we departed from Nature, although it may be convincingly argued, as anthropologist Jared Diamond did, that many indigenous societies also met their demise in the blatant abuse of their natural surroundings – Easter Island comes to mind. Wrecking the place is not a prerogative of the modern mind. It is just that it has many more means at its disposal to do so. Maybe humankind started its disconnection from nature with the advent of agriculture, as is commonly held, bringing in its wake the development of human culture and the concomitant nature/culture split. I am quite sympathetic with that view, I must say. At any rate it is hard to object that our disconnection accelerated with the advent of the Scientific Revolution. We then stopped treating Earth with any form of reverence, as a subject in her own right and with her own agency, and in this realization, I think, lies the message of the *Apollo* mission: we must renew our bond with Earth on the basis of an I-Thou relationship, which recognises Her as a subject without whom we *cannot* become. Some contemporary initiatives such as the Rights of Mother Earth and the concept of Ecocide go a long way towards this recognition but their language is legalistic rather than compassionate. Although coming from the right place their downside is that it defines Nature – a word that, granted, would need further qualification – in our own anthropocentric terms. One has to start somewhere I suppose but I have always struggled with the idea of fighting a destructive capitalist system with its own weapons. As somebody once remarked, environmentalism may have unwittingly done more harm than good: by creating protected natural areas – natural parks, reserves – the unconscious implication was that the rest was up for grabs. And now, even the national parks are under assault. The pinnacle of nonsense, I think, was reached in 2006 with the Stern report, which tried to monetise the services rendered by Nature. To reduce everything to pounds and dollars is the surest path to self-annihilation. The monetisation of nature tries to appeal to mind rather than to heart, and it actually entrenches one into the very destructive mentality that has to be fought. The left hemisphere cannot be reined in by more left hemisphere but by a willing submission to the right, as McGilchrist pointed out. Nothing short of the heart and compassion will do the job.

Deep down I do agree with ecofeminism that the destruction of the Earth is a symptom of a detached Apollonian perspective that rejects the Feminine. Which is where the wanderings of the *Argo* may teach us something. As Edinger highlighted, the wronging of the Feminine by Jason meant that the whole Argonaut enterprise was a failure. All these heroic superhuman efforts ending up in chaos, what terrible waste! I believe that a symbolic reading of the *Apollo* mission leads us straight to the resurgence of the Feminine. For scholar of Indian religion Heinrich Zimmer, the image of the Feminine reappears every time a society needs to dig deep and recover its roots – needs to refresh itself, to drink from the sources of the unconscious when it has become separated from its creative impulse. This was the case in the 12th century when Christianity had become stale and the Church, seen as corrupt, was not carrying strong ethical values any longer. Then the Grail legends came to the fore, with their imagery of the Waste Land and the Wounded King that could only be saved by finding the Grail – the Feminine receptacle. Simultaneously, troubadours and *Minnesingers*[48] were singing the praises of woman, extolling the virtues of courtly love, although Emma Jung and Marie-Louise von Franz cautioned that "woman was only loved externally; the manly ideal was always that of a one-sided and absolute masculinity."[49]

Is Neil Armstrong – carrying the flag for humankind – the new Parzival and if so, which one is he? The Parzival of the first visit to the Grail Castle, who does not ask the question and leaves the Land as wasted as ever? Or the Parzival of the second visit, who shows compassion and is given the Grail, healing the land – "What ails thee, Uncle?" The vantage point from the Moon provides us with a new Archimedean point of reference, and from this position outside our usual earthy plane the Feminine reappears in the guise of Gaia, a blue mandala of wholeness. The picture of Earth shot by Bill Anders aboard *Apollo 8* on Christmas Day vividly embodies this resurrection – *Earthrise* was the name of that picture, a fitting name. At this juncture in our history Gaia is calling us back. We know all too well that it is indeed crucial that we honour Her lest the whole Astronaut enterprise ends up in failure as much as the Argonauts' did.

"What ails thee, Mother?"

Chapter 6

The Moon Landing

The Triumph of Solar Consciousness

It was as if we had begun to turn the pocket of the universe inside out.

– Norman Mailer

The total number of astronauts having walked on the Moon is perplexing: twelve, the solar number, in 6 missions of two astronauts each, from *Apollo 11* to *Apollo 17*. All attempts were successful bar one: *Apollo 13*, the one with the 'unlucky' lunar number – how odd. 'Twelve' immediately calls forth the twelve astrological signs of the zodiac. With some trepidation I looked into the astrological charts of all twelve astronauts. What if each sign of the zodiac was represented on the Moon? That would definitely have tickled the astrologer in me. Short of that, I was expecting fire signs[1] to be the most prevalent among the Sun signs. Just like Edinger, quoted in the introduction of this book, I thought that the archetype of the astronaut was the 'Arian man', the sign of the energetic conqueror. Another strong candidate was Sagittarius, the sign of the explorer. In actual fact, only *one* astronaut had a Sun sign in fire: Neil Armstrong, in Leo,[2] a sign ruled by the Sun. Leo is the sign of the leader, of the King, and the Sun is very comfortable when placed in its own sign. Sometimes too comfortable: a negative expression is an autocratic and selfish tendency, a 'look-at-me' attitude. With the Sun placed at the bottom of his chart, under the horizon, Neil does not show such propensity. His Sun is a midnight Sun – indeed Neil was born at 00.31am in Wapakoneta (Ohio), a name of Shawnee origin meaning 'the place of white bones'. Fittingly, in true Leo fashion, Neil was the first one on the Moon, and if Buzz Aldrin's name is still remembered, few are those who can name any of the other ten visitors to the Moon. For eternity, it is Neil's name that has become attached to the Moon. He has stolen all the limelight, and he has always felt uncomfortable about it – and so, the rumour has it, has Buzz. But Neil never tried to pull the rug in his direction. It just happened: he was the right man at the right time and, again a Leo feature, he had the inner self-confidence to carry out the mission that had been asked of him. As he famously said, he was an engineer through and through, a self-confessed nerd:

> I am, and ever will be, a white-socks, pocket-protector, nerdy engineer – born under the second law of thermodynamics, steeped in the steam tables, in love with free-body diagrams, transformed by Laplace, and propelled by compressible flow. As an engineer, I take a substantial amount of pride in the accomplishments of my profession.[3]

This statement is truly remarkable. The opposite sign of Leo is Aquarius and while Leo is fiercely individual, Aquarius subsumes the individual into the collective. Opposite signs are often at odds with each other and offer uncompromising reflections of each other's shadows. But to have 'pride in the accomplishments of [one's] profession': this is Leo having well integrated his Aquarius opposite, and I think this sums up Neil's personality very well. He basked in the light of the profession, not the other way around. Neil was at pains to point out that the Moon landing owed more to engineering than to science. But, with his tendency to hide in the collective, a very reluctant hero he was. One of the few astronauts not to have written an autobiography, he was hiding from the public eye as much as his fame allowed him to. His Sun in Leo was placed in the most private part of the chart – a solid foundation to draw on, an unshakable self-confidence, but a very modest one too. An innate sense of destiny, without ostentation, unapologetic but deeply uncomfortable with the exposure. Neil's relation with his father would no doubt shed further light: with the Sun in this position, any astrologer worth her salt would be curious to know a bit more about that.

Buzz stepped on the Moon twenty minutes or so after Neil. He will always be known as the second man on the Moon, although he had the "dubious distinction", as he himself put it, of being the first one to pee on the Moon. Crumbs of recognition are hard to get by, although being part of the first mission to the Moon should have been enough to satiate anybody's appetite. Michael Collins, who stayed in Command Module all along and circled the Moon while the other two were catching all the attention, had no such qualms about not being the first one, nor the second, nor anything at all in fact since he never made it to the surface. Collins was blessed with a cheerful character; "Collins was cool." said Norman Mailer in *Moonfire*, his account of the *Apollo 11* mission. "[He] was the man nearly everybody was glad to see at a party, for he was the living spirit of good and graceful manners."[4] Buzz had a much more complex personality. He did not get along with his father, and his mother Marion, like her own father had done before her, committed suicide in May 1968. She was apparently struggling with her son's budding fame as an astronaut. What makes this tragic story all the more baffling is that Marion's maiden name was Moon. You cannot make up things like that. This, truly, is dumbfounding: the mother of one of the two astronauts who first walked on the Moon, a woman whose name was Moon, took her own life one year before the Moon landing, an event that itself can be seen as the death of the Moon.

Suicide inevitably brings very disturbing emotions to the surface. When I was locked into my cell in Paris, my personal Moon landing, I did contemplate my own death with frighteningly open eyes. All those years later, I still shiver with the thought of it. At one point – probably in the middle of the second day – I had reached utter despair. I think the worst for me was the prospect of being unable to face the people I loved – my wife, my children, my friends and family – the feeling of having failed them, the impossibility to look them in the eyes ever again, the absolute shame. This sounds so incredibly overdramatic today. I was so naïve, so tender. But with all these thoughts running amok in my little panicky brain, desperate for an escape route, I was considering the different options. Not that I was going to act on the spot, no – rather, I was preparing for a worst-case scenario, for the possibility of things going really awry in the months ahead. It was a very theoretical exercise, I admit it, whose goal was maybe no more than lifting the lid on this unbearable feeling of being trapped. To recover a modicum of agency, a sense of being in charge. I did not want to go violently, though, and eventually I settled for what seemed the best option. I love the mountains and I felt that going under the stars, in an open night sky, in the fresh and purifying air of the mountains, was the best way to hang the hat. I would go to Chamonix and use the cable car that rides up to Aiguille du Midi, 3,800m high. From there on I would hike on the glacier, find a sheltered spot underneath one of the giant peaks towering all around, and swallow a whole bunch of sleeping pills. There. Peacefully die of cold in my sleep, like a light gently extinguished, a little star reuniting with her big sisters above. Notwithstanding Dylan Thomas, I would go gently into the night.

I admit that this is pure speculation on my part but with such a chequered family background Buzz probably craved recognition, which may explain his queasiness about his number 2 role. The relationship between Neil and Buzz was purely professional, and by all accounts, there was no lost love between them. After the Moon landing and the harassing public engagements, they hardly got in touch ever again, even though one would assume that sharing such an intensely successful human adventure would have created lifelong bonds. The way I see it, if Armstrong was the Sun, Aldrin was the Moon, a polarity that they never managed to bridge – no sacred marriage between them. In the astrological chart, the Moon represents the actual mother, but also the sense of security, safety and nourishment that one's mother is supposed to provide. It also represents emotional life. In Buzz's chart, the Moon is harmoniously placed in Libra but is in hard aspect to Pluto, the Lord of the Underworld – an intense aspect that flies in the face of the comfort and security that the Moon craves. Buzz was an emotional man, in contrast to Neil's calm, composed demeanour. "There was… a special depth to the 'curious mixture' called Buzz Aldrin,"[5] said Farmer and Hamblin in *First on the Moon* – a "magnificent confidence, bordering on conceit, and humility,"[6] as his first wife Joan once remarked. There was also a very strong religious bent to Buzz, one that he was not shy to express. Hamblin recalls how Buzz's personal kit on Gemini 12 "had included a prayer in the handwriting of his boyhood nurse Alice" – a touching and revealing emotional nod to his old caretaker:

The light of God surrounds me
The love of God enfolds me
The power of God protects me
The presence of God watches over me
Wherever I am, God is.[7]

"Beautiful view," said Buzz while stepping off the ladder on to the lunar surface: a perfect illustration of Moon in Libra. "Isn't that something!" replied Neil. "Magnificent sight out here." Buzz then recounts:

I slowly allowed my eyes to drink in the unusual majesty of the moon. In its starkness and monochromatic hues, it was indeed beautiful. But it was a different sort of beauty than I had ever before seen. *Magnificent*, I thought, then said, "Magnificent desolation." It was a spontaneous utterance, an oxymoron that would take on ever-deeper dimensions of meaning in describing this strange environment.[8]

While Neil would forever be remembered for his small step and mankind's giant leap, Buzz settled for 'magnificent desolation' – Pluto in square to the Moon enters the scene, immediately adding an element of dread to the aesthetic Libra Moon. Aldrin would use this signature statement for the title of his 2009 autobiography, for indeed the "oxymoron would take on ever-deeper dimensions of meaning". Buzz was 39 in 1969 and, as he poignantly remarked in his autobiography: "What does a man do for an encore after walking on the moon?"[9] What is left to do, and how to give meaning to such an extraordinary adventure?

I was discovering that everyone wanted me to answer the inevitable question: 'what is it like, being on the moon?' I struggled with an answer. I wanted to say something profound, something meaningful. But I was an engineer, not a poet; as much as I grappled with the quintessential questions of life, questions of origin, purpose, and meaning… I found no adequate words to express what I had experienced… While on the surface of the moon… [I] had called it 'magnificent desolation'. Now these words seemed to describe my own inner turmoil as I thought about the days ahead.[10]

In the days, and years, ahead, Buzz struggled with depression and alcoholism. Depression is a signature of Pluto square Moon, the terrible *angst* of living, the difficulty to find peace and quiet, to rest and feel safe. There is a price to pay for desecrating the Moon. Neil became a recluse and Buzz an alcoholic. A kind of curse of the pharaohs, although both of them may be said to have bounced back from the straining aftermath of the mission. My persistent reference to astrology would no doubt irk these hardened engineers, who probably had no time for such nonsense. But the beauty of astrology is its extraordinary resilience, sometimes cropping up in the most unexpected places. Whilst on the way to the Moon, the Control Centre in Houston informed the three astronauts that "a Houston astrologer, Ruby Graham, says that all the signs are right for your trip to the moon. She says that Neil is clever, Mike has good judgment, and Buzz can work out intricate problems... Buzz is said to be very sociable and cannot bear to be alone."[11] To which Michael Collins replied: "Who said all that?", while laughter could be heard in the spacecraft. On the face of it, another example of left brain sneering but still... why mention the astrologer in the first place – who, credit to her, got it right? Is it really just harmless fun or is something lingering beneath the surface, the remnant of old superstition, a disavowed need to believe? A while later, while Collins was all alone circling the Moon aboard *Columbia*, his wife Pat laughed when she heard him say that he had seen a bright light. "It's Venus, Mike," she said. "He always sees Venus. It must be in his horoscope."[12] Well, spot on, Pat. In Michael's chart, Venus is in a very prominent position, at the apex of an important configuration called a T-square. She is also contacting all the other planets bar Mercury, the only planet to do so. She symbolises his "graceful manners".

Michael may always have seen Venus but, as ill-fated *Apollo 13* commander Jim Lovell once remarked, this did not help him navigate the stars:

> For spacecraft navigation... we use thirty-seven stars – plus the earth, the sun and the moon. We don't use Venus, but we do use Polaris, Rigel, Capella, Canopus, Sirius, Antares, Vega, Arcturus, Altair and a big one called Fomalhaut which is less widely known.[13]

With their Latin or Arabic roots, I find star names infinitely enchanting. Fomalhaut, an Arabic name that means 'the fish's mouth', is the brightest star of Pisces Austrinus, a constellation located in the southern hemisphere, also called the Southern Fish. This constellation merges with Pisces, the sign of the two fishes swimming in opposite directions. To my great surprise Pisces turned out to be the most prevalent astrological sign among the twelve astronauts who walked on the Moon – three of them were of the sign, a ratio of 25% against a mean of 8.33%. I am no statistician but it has to be very significant. Then I found that Yuri Gagarin, the first man in orbit, was a Pisces too, and so were three of the Mercury Seven. What is going on?

One of the most remarkable works of Jung is *Aion*, written when he was 76, at a time when he had decided not to shackle himself any longer with the pursuit of academic and scientific respectability. In the foreword to *Aion*, Jung explained that "my investigation seeks, with the help of Christian, Gnostic and alchemical symbols of the self, to throw light on the change of psychic situation within the 'Christian aeon'."[14] *Aion* is an exploration of the archetype of the Self that Jung thought Christ embodied. It is also a groundbreaking piece of astrological work. Born at the beginning of the Age of Pisces, Christ and Christianity have always been associated with Pisces symbolism, and for Jung, there is little doubt that the early Christians, well-versed into astrology, deliberately associated Christ with the nascent aeon: "through the fish symbolism, Christ was assimilated into a world of ideas that seems far removed from the gospels – a world of pagan origin, saturated with astrological beliefs to an extent that we can scarcely imagine today."[15] Christ was the fish – *Ichthys* – and the apostles, once fishermen, became fishers of men. But Christ, as the Lamb of God, was also symbolically associated with the finishing Age of Aries, the Ram, thereby bridging the two aeons. In another significant piece of symbolism, Christ was born seven years after the great conjunction of

Jupiter and Saturn in Pisces, a conjunction that Jewish prophets had said would herald the coming of a Messiah. The stage was set for the next phase in consciousness, and Jung proceeds in *Aion* to explore the psychological implication of this Christian unfolding: as Maggie Hyde noted in *Jung and Astrology*, Jung saw in "the duality inherent in the fish of Pisces [a reflection of] Christianity's irreconcilable opposites."[16] Christ is an all-light figure and needs an Anti-Christ, a Satan opposite that carries the projection of darkness: "the Christ-symbol lacks wholeness in the modern psychological sense," said Jung, "since it does not include the dark side of things but specifically excludes it in the form of a Luciferian opponent."[17] This polarization is typical of a solar consciousness that cannot reconcile the opposites. And since God cannot be evil, or create evil, then evil is an all-human prerogative, a refusal to follow the word of God, an absence of good (*privatio boni*). Jung noted how the Jupiter/Saturn conjunction was opposed by Mars; he saw in it a symbol of the demonization of the instincts: "the planet correlated with the instincts [Mars] stood in a hostile relationship to [the conjunction], which is peculiarly characteristic of Christianity."[18] However, the most remarkable achievement of *Aion*, astrologically speaking, was the way Jung mapped the evolution of Christianity and of Western history by following the precession of the equinox through the vast constellation of Pisces: a *tour de force*, "this work is of the highest order of astrological poetry," says Maggie Hyde, "with its blend of technical skill and symbolic sense, its depth, its scale and scope and the inspiration it offers."[19]

Are the astronauts the embodiment of Christ consciousness, then? In astrological lore, Pisces exhibits a very elusive nature, a craving to reconnect to the Source, a longing for dissolution. *Apollo 17's* Eugene Cernan, the last man on the Moon, had both luminaries, Sun and Moon, in Pisces. In space he felt "wrapped in infinity", feeling that he was about to "touch the face of God."[20] One could not better express Piscean sensibility. *Apollo 15's* Jim

Irwin, Pisces Sun, felt the presence of God while looking for rocks on the surface. There may be something of that longing in many astronauts, an unconscious desire to meet the Creator who, in our Christian heritage, is out there, or up there, but usually not down here on Earth. A sort of idealization of life may also be a by-product of floating high above, and Pisces is a very idealistic sign. In space, as many astronauts have reported, terrestrial frontiers do not exist, conflicts are not visible anymore, our human frailties disappear in the encompassing glow of Earth, her radiating and soothing light. All is peace and quiet, bathed in cosmic consciousness. And then there is the imagery of Pisces itself, ruled by Neptune – the sea, sailors, boats, navigation – that is found in space travel: star sailors (astronauts), spaceship, the image of 'floating on the dark ocean of space'. A natural affinity between space and sea, then, and between Pisces and astronauts. The poetry, to me, is unmistakable. This is not to say that all astronauts had similar experiences – *Apollo 8's* Bill Anders, Libra Sun, actually *lost* his faith while in space.

Are current astronauts still predominantly Pisces, however? I do not know and they may very well not be. One reason might be that space travel may have become a bit more technical, less idealistic, more like a normal job than it used to be when the pioneers were in search of a new territory, exploring physical frontiers that, in many ways, represented the frontiers of consciousness. Maybe astronauts, then, were unwittingly psychonauts too, and may no longer be. But in so saying I may be coming dangerously close to literalism. One should give breathing space to the symbol and not twist its arm. I am not pretending to have come up with a new scientific law – the Sibaud theorem – that says that 'astronauts are majoritarily Piscean'. All I am saying is that when I looked this is what I found and it was symbolically relevant. I do not exclude the possibility that I have been so sensitive to Pisces symbolism among the astronauts precisely because I am a Pisces myself – the hermeneutic loop, raising its diabolical head yet again. If I had looked at another time, under different circumstances, I may have found something else, and that something else might have been just as relevant. Symbolism operates in the here-and-now – *hic et nunc*:[21] the oracle speaks and what she says is relevant for this time, for this place and for *you*. If you were to ask again – not advisable – you would get a different answer. That is what is so incredibly difficult to comprehend for the left hemisphere, which can only think in general laws and dismisses the particular – variously referred to as chance, coincidence, placebo, anomaly, exception to the rule, etc. A symbolic attitude does precisely the opposite: it is what stands out of the ordinary that is meaningful. This was the *modus operandi* of traditional indigenous societies, which were paying a lot of attention to any event occurring outside the normal course of day-to-day life. One should therefore not be surprised that the symbol just *cannot* be accommodated by the left brain, which simply operates alongside a different spectrum.

The foray into Fomalhaut has taken us into a hell of a tangent. I want now to revert to Jim Lovell's quote about the navigational stars, focusing in particular on to one of them, Rigel, which will lead us into Orion. Rigel is Beta Orion, the second brightest star of the constellation, and is located on the left toe of the hunter. For astrologer Bernadette Brady, who has done a lot of research on fixed stars, "in ancient Egypt, to be under a pharaoh's foot symbolised being under Osiris' protection."[22] Orion is of particular interest because it features prominently on the Apollo programme patch.[23] The central character of the patch is the letter A, after Apollo, and the three stars of the Orion belt – Mintaka, Alnilam and Alnitak – form the bar of the letter, a neat counterpoint to the three astronauts of the spacecraft. On either side of the letter is Earth, on the right, and the Moon, on the left, a Moon, however, drawn with the colour of the Sun and with Apollo's face – a clear statement of intent. The main other Orion stars are represented: tangential to the upper side of the Moon is the red star Betelgeuse – Orion's shoulder, Alpha Orion – while Rigel is located at the lower right of letter A, with Saiph and Bellatrix at their respective places in the constellation. Orion's sword is drawn between the legs of letter A, with the Orion nebula at the centre of the sword. There are a number of interesting pieces of symbolism here. Brady notes that "many modern theologians/mythologists agree that [the] precessional movement of Orion is the source of the religious concept of an afterlife given to us by a god who dies. Osiris/Orion is a 'prefiguration of Christ'."[24] We are thrown back into Piscean and Dionysian territory. The neat conjunction of Betelgeuse with the Moon, drawn on the patch, is puzzling. For Brady, again, "the planet in contact with Betelgeuse will represent talents or abilities... which can be used for joy, success, or even fame... Whatever [Betelgeuse] touches, it will produce positive results."[25] It certainly did with the Moon landing.[26] The myth of Orion itself is very aptly chosen,[27] bringing as it does Apollo/Sun and Artemis/Moon together, albeit in rather gruesome fashion. According to Robert Graves,[28] one version of the myth tells how the hunter Orion, the most handsome man alive, boasted that he could kill any beast or monster that he would meet. Apollo, jealous of his sister's attraction to the hunter – with whom she shared the passion of the hunt – went to Mother Earth who, on hearing Orion's boast, sent a giant scorpion to pursue him. Neither Orion's arrows nor his sword could get rid of the scorpion, so Orion dived into the sea, attempting to reach Delos for protection. Apollo then called his sister Artemis and tricked her into believing that the head of the hunter, far away at sea, was that of a man who had seduced one of her Hyperborean priestesses. Angry, Artemis took aim and shot Orion through the head. Realizing her mistake, desperate, she set Orion into the stars, where he has been ever since chased by Scorpio, the scorpion – the constellation directly opposite Orion, which borders Taurus.

When one patches these different pieces of symbolism together, it is hard not to be mystified by NASA's esoteric proficiency. Add Egyptian symbolism, and the conspiracy theory mill is turning at full speed again. The Egyptian New Year was said to start at the heliacal rising of the star Sirius at Heliopolis, a day that was assessed in 139 CE by the Roman writer Censorinus as 20 July. *Apollo 11* left Earth on 16 July and landed on the Moon on 20 July. Coincidence? Masonry is ripe with Egyptian symbolism. Equally baffling is the fact that 16 July was the first day of the Mayan calendar year, the *Haab*. The connection between pre-Hispanic Mexico and Egypt is troubling. At Teotihuacan, the pyramids of the Moon, of the Sun and of Quetzalcoatl, built in the 2nd and 3rd centuries CE, were positioned in a similar way to the three pyramids of Giza, built three thousand years before – all of them mirroring the position of the three stars of Orion's belt. Besides, the Mayas dated the birth of their mythical fourth Sun on the (Gregorian-equivalent) date of 13 August, 3114 BCE[29] – astonishingly, it was at this approximate time that the Lower and Upper Kingdoms of Egypt were united for the first time, birthing the first dynasty. Was magic invoked to facilitate success on the Moon? It is not the place here to answer these questions – I am in no position to, anyway. It is enough to raise them. There seems to have been more than meets the eye in the lunar venture, an occult dimension that no amount of scientific or technological speech can entirely conceal. Or am I foolishly following Alice in her journey to the land of the grinning cats?

The astrology of the mission yields very interesting clues as well, and I will highlight two salient features. On departure day and time (16 July at 09.32am Eastern Daylight Time from Cape Canaveral) the Sun was in the Moon sign (Cancer, ruled by the Moon) while the Moon was in the Sun sign (Leo, ruled by the Sun). Two planets in each other's signs are said to be in mutual reception – an astrological signature that emphasizes a harmonious relation between the two planets involved. Sun and

Moon in mutual reception makes a lot of sense when you want the God of the Sun to visit the Moon – he is welcome there. If we now move to the chart of the actual landing (20 July at 4.18 EDT, calculated for Houston) we find the Moon at 7 degrees 53 minutes Libra. I compared this chart to the natal chart of NASA itself (1 October 1958 in Washington, DC, calculated for 12pm). In NASA's natal chart, the Sun is at 7 degrees 58 minutes Libra. The time of 12pm is an approximation but the astrological position of the Sun will not change much during the day. So, we are basically left with the remarkable fact that the Moon of the Moon landing horoscope is perfectly aligned with NASA's natal Sun.[30] The Sun in NASA's chart represents NASA's mission, its essence, its profound expression, *where it shines*. NASA was created to respond to the Soviet challenge raised by Sputnik. It was to establish the primacy of the United States in space and this objective quickly became symbolised by achieving a Moon landing and back. One would be hard-pressed to find a more suitable symbol: the astrological Moon of the actual Moon landing is aligned with the natal astrological Sun of an organisation whose *raison d'être* was to land on the Moon. At this point we are left with a chicken and egg conundrum. Does this piece of symbolism highlight the mysterious ways in which astrology is working – which is a bad way to put it, really, because astrology does not 'work' but reflects? Or was there a deliberate intention to land on the Moon at this time so as to ensure success? Did NASA work with astrologers to elect a favourable time?[31] One has to realize the extreme technical difficulty to do so. The Moon moves fast in the sky (13 degrees a day) so to ensure a conjunction with NASA's Sun,[32] NASA would have had to time the landing within a range of four hours, which, given the amount of potential technical glitches, seems very ambitious…. But maybe it is possible, I don't know. It may in fact be utterly naïve on my part to imagine that NASA had not planned anything to the second. As recounted on its own website:

> The descent engine continued to provide braking thrust until about 102 hours, 45 minutes into the mission. Partially piloted manually by Armstrong, the Eagle landed in the Sea of Tranquility in Site 2 at 0 degrees, 41 minutes, 15 seconds north latitude and 23 degrees, 26 minutes east longitude. This was about four miles downrange from the predicted touchdown point and occurred almost one-and-a-half minutes earlier than scheduled. It included a powered descent that ran a mere nominal 40 seconds longer than preflight planning due to translation maneuvers to avoid a crater during the final phase of landing.[33]

So the Eagle landed "one-and-a-half minutes earlier than scheduled" – than scheduled when? Before lift-off or much later, once the descent to the lunar surface was initiated? In *First on the Moon*, Farmer and Hamblin describe the arrival in lunar orbit:

> It was only the first revolution, the first of twelve around the moon, before the command and service module was due to separate from the lunar module during the thirteenth pass, on the back side of the moon. But a big decision had been taken: it was go.[34]

Twelve revolutions around the Moon before initiating descent on the thirteenth. The symbolism is screaming again but I rather want to focus on the decision-making process. Was there always going to be twelve revolutions or was it left to the astronauts to estimate when the descent could proceed? That truly is rocket science! We are faced here, in my view, with two equally fascinating options: either NASA could and did plan the landing consistent with a Moon at 7 to 9 degrees Libra, or the landing happened to take place at this remarkable symbolic time.[35] The first scenario proves NASA's esoteric leanings, the second takes us right back to the enigma that is astrology. In both cases, a defeat for the left brain! Unless of course it is all dismissed as mere chance, just a meaningless coincidence, the usual dump bag.

But maybe now is an opportune time to dip our toes into the mystery of astrology. For indeed, any practitioner, or anyone who has taken the time to study the subject with open mind and heart, cannot fail to be bewildered by these 'coincidences' – namely, the uncanny propensity of the symbol to crop up unannounced, in a way that is filling one with wonder and awe, very much like a poetic image suddenly pervades one's whole being. Impromptu, the symbol unexpectedly reveals a hidden message, unveils a secret scripture behind the canvas, and one is both mesmerised and humbled. This is proper ecstasy, in the Greek sense: one is taken out of oneself, and suddenly attunes to something very big and very mysterious indeed. NASA's Sun is exactly on the Moon landing's Moon? Try as I might, I just cannot dismiss this as irrelevant, as meaningless coincidence. It points to something, and maybe what it points to is no more than my own obsession with Sun/Moon symbolism! Maybe I am caught in the hermeneutic loop, in fact of course I *am* caught in the hermeneutic loop, but is that really all there is? Anyone engaging with the symbol has to navigate these tricky waters, for Hermes is the consummate trickster. And the elusive God may be playing a trick on the practitioners of astrology themselves: for while astrologers constantly play with astrological symbolism, they very rarely engage with the symbol that is astrology itself – in fact they may often be said to literalise astrology, falling into the trap of objectivity.[36] Astrology is not a science of the Heavens, it is an artistic expression of the Human as Two, as Kripal would call it. Astrology points us back to us – a wink from the dark.

Dean Woodruff was no astrologer, but he knew the

importance of a symbol. Woodruff was a Reverend, the minister of the Webster Presbyterian Church that Buzz Aldrin attended in Houston and in which Buzz was an elder. Shortly before the Moon mission, Woodruff and Aldrin met up to discuss how to give a more symbolic dimension to the historic trip to the Moon. The result was a piece of writing by Woodruff entitled *The Myth of Apollo 11: The Effects of the Lunar Landing on the Mythic Dimension of Man*. This is a remarkable text. Woodruff starts by emphasizing the human dimension of the venture:

> What happens today is an expression of man's ability – of man's self-determination. It is the molding of knowledge and theory; it is the channelling of human resources in solving problems; it is dreaming dreams and having visions; it is the concretizing of man's potential.[37]

The "molding of knowledge and theory" is symbolised in the Moon landing horoscope by the triple conjunction[38] of Jupiter (knowledge, expansion), Uranus (theory, technology) and Moon. Woodruff then moves on to Nietzsche's Übermensch: "it is not a new biological species but a new kind of man who realizes his capacity for self-transcendence and self-fulfillment... This is what Nietzsche developed in his idea of the 'superman'."[39] The Reverend goes even further in extolling the virtues of Man by taking him dangerously close to God: "Today, Armstrong, Aldrin and Collins,... as 'representative man', will implicitly ask the question, 'What is man... thou hast made him little less than God, and dost crown him with glory and honor'."[40] Throughout his paper, Woodruff leaves no doubt as to the historical context of the Apollo mission:

> Since World War II we are in the advent of the modern worldwide civilization that is based upon science – for the language of science and technique is the same in every

country. If we are, in fact, in the midst of this new worldwide civilization, then the first symbolic event of that civilization is 'the bomb'. The second symbol event of that civilization is Apollo 11, the first and most imaginative non-destructive event of a new civilization.[41]

Woodruff contrasts 'Little Boy' with *Apollo 11* but the background is the same, and it takes us right back to the solar consciousness that gave birth to science and technology. Apollo rules, even if his image has been seriously distorted by centuries of Western left brain *modus operandi*. The *Apollo* landing on the Moon can be seen as the literal enactment of the slaying of the dragon at Delphi. It is the tip of the arrow, the pinnacle of a process started four thousand years ago in Babylon. Neil Armstrong is the flag-bearer of a solar consciousness that Zen Buddhist scholar DT Suzuki saw etched on the very face of Western man:

The West... has a pair of sharp, penetrating eyes... which survey the outside world as do those of an eagle soaring high in the sky... And then his high nose, his thin lips, and his general facia contour – all suggest a highly developed intellectuality and a readiness to act. This readiness is comparable to that of the lion. Indeed the lion and the eagle are the symbols of the West.[42]

The griffin, a mythical animal, reconciled both solar animals by combining a lion's body with eagle wings and head – and the griffin was one of Apollo's animals. French historian Michel Pastoureau extensively researched animal symbolism in the Middle Ages and noted the overwhelming presence of the lion in the imagery of medieval coats of arms – the lion was present in 15% of them, with the eagle in second place with 3%.[43] From the 11th century onwards, the lion became associated with all sorts of qualities and virtues: courage, strength, pride, justice, largesse. The lion became the emblem of the Christian Knight, opposed to the dark Dragon of the Pagan. This attitude to the lion, says Pastoureau, represented a remarkable change of heart. In the 5th century, Augustine was fiercely opposed to the lion, in which he saw an embodiment of violence, cruelty and tyranny.

Among the Northern Europeans and the Celts, the lion was ignored in favour of the bear, an animal obviously much closer to home. Bear was 'Art', as in Arthur, and was Artemis' animal. Pastoureau highlights the role of the Church in this turnaround. At the beginning of the second millennium the lion acquired a Christological dimension while simultaneously a double move was orchestrated: to embody the negative qualities of the lion, the image of the leopard appeared – a negative lion, with most of the features of the lion except for the mane. The leopard was evil, the Anti-Christ. We may note that the leopard was the animal of Dionysus who, despite being an archetype for the resurrected nature of Christ, was no friend of the Church. And furthermore, the bear became demoted. The bear had a violent nature, was lubricious, a hairy version of man, bestial, and – an unforgivable crime – was indigenous to Europe and symbolised the forest, the hidden forces of paganism. The Church relentlessly chased the bear, which then became an object of fun, domesticated, tortured, shown around in fairs like a grotesque, diminished man, a fool. A similar movement, says Pastoureau, took place between the boar and the stag. Initially revered and admired for its courage and valour, the boar was gradually demonized by the Church for its base instincts, and replaced by the stag, a solar, Christ-like animal. One way to demonize the boar was to debase its hunt. From the 12th century onward, the Kings no longer hunted the boar but the stag – which became the noble, royal hunt. A famous legend spoke of Eustachius, a Roman soldier and keen hunter, who saw the sign of the cross appear between the horns of a hart he was pursuing and converted to Christianity on the spot. Forget the sexual symbolism of the animal: the stag was gradually clothed with the attributes of a pure and virtuous animal, like the lamb or the unicorn. The stag was a sacrificial animal, put to death ritualistically; this ritual death symbolised Christ's Passion. Its name itself – *cervus* – was likened to *servus* – Christ, the Saviour.[44] The boar, the animal of the Goddess, was relegated to the deep confines of the forest, a devilish and lascivious beast banished from the company of virtuous men.

The lion and the eagle, 'the symbols of the West', are both present on the Moon. Neil Armstrong, Sun in Leo – a sign ruled by the Sun whose symbol is the lion – is the latest avatar in a long lineage of solar heroes whose mythical father was Marduk. In Armstrong's 1968/1969 solar return chart,[45] the Sun in Leo is about to rise, conjunct the Ascendant in Leo in the first house. It is dawn, just before sunrise, the promise of a bright new day. One would expect to shine in that year, and that light would be for everyone to see. A similar horoscope can be cast for the Moon, called a lunar return chart: it is the chart calculated for the moment when the Moon returns to her position in the natal chart, every month. In Armstrong's lunar return chart cast for the month of the Moon landing, the Moon has just risen – she is on the Ascendant in Sagittarius in the twelfth house. Sun on the Ascendant in the solar return chart, Moon on the Ascendant in the lunar return chart: the astrology of Neil Armstrong for the Moon landing was staggeringly meaningful. Armstrong, Leo by birth, Sun conjunct Ascendant in the solar return chart, in command of *Apollo*, a mission that landed on a Sun-day – the lunar venture was literally oozing with Sun symbolism.

And then there is the eagle, the bird of Zeus, the solar bird *par excellence*. The eagle loomed large over the mission, conspicuously drawn on the patch of *Apollo 11*,[46] holding an olive branch in its claws while landing on the lunar surface, with Earth in the background. NASA knew the importance of symbolism: a major question for the astronauts was to find call signs for the CSM (Command Module) and the LM (Lunar Module). As Hamblin remarks in *First Mission*, "there was considerable sensitivity in the astronaut community about the choice of call signs for the first lunar landing."[47] Armstrong, Aldrin and Collins received hundreds of suggestions and as late as April 1969 they were still debating:

> The three Apollo astronauts – and their wives – were still rolling 'paired' names off their tongues for sound effects: Romeo and Juliet; Anthony and Cleopatra; Daphnis and Chloe, the Greek shepherd and shepherdess (Daphne herself was brought up but rejected because she spent half her time trying to escape from the god Apollo).[48]

One anecdote is particularly revealing: in 1967 or 1968, astronaut Pete Conrad, who was then touted to take charge of the mission, was looking at potential call names when someone suggested 'Venus': "Conrad liked that, looked up Venus in an encyclopedia and announced: 'She won't do.' As patroness of Pompeii Venus had somehow got herself associated with prostitution."[49] What a sad state of affairs when the Goddess of Love and Beauty cannot evoke anything else but the rejected Feminine, the exiled and demonized Mary Magdalene of the Gospels.[50]

By mid-1969, "the astronauts," said Farmer and Hamblin, "were sure in their minds that they wanted the names to reflect a degree of American pride, but within bounds set by taste and dignity."[51] Simultaneously, the astronauts had also been charged by NASA to design the patch of the mission. "We wanted to keep our three names off it," explained Collins on the NASA website,

> because we wanted the design to be representative of everyone who had worked toward a lunar landing, and there were thousands who could take a proprietary interest in it, yet who would never see their names woven into the fabric of a patch. Further, we wanted the design to be symbolic rather

than explicit.[52]

It was astronaut Jim Lovell who suggested the eagle, the national bird of the United States, as the focus of the patch. Tom Wilson, a computer expert and the *Apollo 11* simulator instructor, then proposed that the eagle carry an olive branch "as a symbol of the peaceful nature of the mission."[53] Initially positioned in the bird's beak, the olive branch was moved to its talons when it was judged that the empty extended talons of the bird were too hostile and warlike:

> Highly realistic, the crater-pocked moon was colored grey, the eagle brown and white, the Earth blue, and the sky black (just as it would be from the lunar surface). The Earth, suspended like a small blue marble in a black sky, is actually incorrectly drawn. The patch shows the Earth to be shadowed on the left side, while the Earth, if viewed from the lunar surface, would be dark on the bottom. This mistake was never corrected.[54]

That is odd. How could NASA, which prides itself on its excellence and professionalism, let slip such a mistake? Is there a hidden significance to this mishap? I don't know, one of course can quickly get carried away with the pull of the occult. Conscious or not, this skewed representation of Earth is disturbing: I see it as the symbol of our neglect for Her, of our lack of care and respect, of our casualness. Through that lens, the Earth is almost an afterthought, an irrelevance, a distraction in the heroic journey onwards and upwards – why even bother represent Her correctly? On the patch, all eyes are on the conquering eagle, the vanguard of our destiny.

"After the patch design had been approved," says Hamblin, "the problem started to solve itself. 'Eagle' for the lunar module came straight off the patch."[55] Columbia was chosen as the name for the CSM: "Columbia was also a national symbol," said

Armstrong, "but more importantly the choice was an attempt to reflect the sense of adventure and exploration and seriousness with which Columbus undertook his assignment in 1492."[56] The space conquistadors indeed made it to the other shore. On landing, Armstrong coolly spoke the first words ever uttered on the Moon: "Houston. Tranquility Base here. The *Eagle* has landed."[57]

The Eagle has landed. Solar consciousness had taken over. Conquering Moon, escaping Earth – the Apollonian dream of a soulless consciousness that looks up to the stars for redemption. The consciousness inherited from Ancient Greece found its highest expression, its apogee, on the surface of the Moon. As Joseph Campbell put it:

> When the [Greeks] kissed their fingers at the moon... they did not fall on their faces before it, but approached it, man to man, or man to goddess – and what they found was already what we have found: that all is indeed wonderful, yet submissive to examination.[58]

The Moon Dragon had been well and truly defeated by the solar rays – the arrows – of an Apollonian consciousness that literally trampled on her body. Her defeat is the defeat of the Goddess, of all goddesses. For, as much as the conquest of the Moon, the *Apollo* landing also symbolised man's desperate struggle to free himself from the constraints of nature, as expressed by Armstrong: "the important achievement of *Apollo* was demonstrating that humanity is not forever chained to... planet [Earth]."[59] One discerns in these words the echoes of the masculine quest to escape from the clutches of the All-Devouring Mother – "the tunnel vision... so focused on conquering the unconscious mother," said Marion Woodman.[60] Buzz Aldrin was even cruder:

We can continue to try and clean the gutters all over the world and spend all of our resources just looking at the dirty spots and trying to make them clean. Or we can lift our eyes up and look into the stars and move forward in an evolutionary way.[61]

The terror of being chained to Earth, of being her prisoner, runs deep. It is expressed in the current transhumanist movement, whose article of faith proclaims: "Humanity stands to be profoundly affected by science and technology in the future. We envision the possibility of broadening human potential by overcoming aging, cognitive shortcomings, involuntary suffering and our confinement to planet Earth."[62] In a similar fashion, the Reverend Woodruff quoted Eliade and 'the bondage to earth':

[The magic flight] is found everywhere, and in the most archaic of cultural strata... the longing to break the ties that hold him in bondage to the earth is not a result of cosmic pressures or of economic insecurity – it is constitutive of man... Such a desire to free himself from his limitation, which he feels to be a kind of degradation... must be ranked among the specific marks of man.[63]

I have real difficulty reconciling myself with Eliade's take on the magic flight. For Eliade, "the most representative mystical experience of the archaic societies, that of shamanism, betrays the *Nostalgia for Paradise*, the desire to recover the state of freedom and beatitude before 'the Fall', the will to restore communication between Earth and Heaven."[64] This sounds to me like shamanism looked through a very Christian lens. As far as I can see, shamans are the most deeply rooted people on the planet, the most attuned to Earth herself. They are in a deep inner dialogue with fire, rocks and water, and with the animals –

some of them, their power animals and guides, more than others. I just cannot see the magic flight as an attempt to escape from the 'bondage of the earth'. It is the opposite, in fact: shamans do indeed access other dimensions than the material world but they do so in order to receive boons for the benefit of the community and/or of the individual. Crucially, *shamans engage with the Earth as a Thou*. "A kind of degradation," says Eliade – to be rooted in Earth, that is? This, truly, is the negation of lunar consciousness: the imposition of a solar agenda that refuses to recognise death as part of life, and that is locked in a desperate struggle to free itself from what it perceives as the shackles of biological life – *bios* – and is oblivious to *zoe*, the higher principle. A refusal to pay the price of being part of the web of life, a fantasy of escapism, as if life on Mars, Alpha Centauri or the Orion Nebula was promising endless felicity, a blank cheque free of death and suffering. A consciousness stuck in a dialogue between It and It. This is the stance of the soulless wanderer, either negating Spirit, or so desperate in his quest for it – for the Sun – that he forgets his kinship with the rest of Creation. And the more he estranges himself from Her, the more desperate and destructive the quest, the blinder the faith – in a transcendent and omnipotent God, in the rising rocket, in benevolent aliens, in virtual reality.

At the end of his text, the Reverend Woodruff steps into the symbolic realm, acknowledging the cruel lack of myth in our technological world – something that Jung and Campbell had pointed out before him:

Science, as the achievement of man, has created a worldwide technical civilization and, as yet, has not given birth to any cultural symbols by which man can live. The Apollo event comes at a time when we need a symbol, and need to tap a myth that will graphically express the unending journey outward. Perhaps when those pioneers step on another planet and view the earth from a physically transcendent stance, we can sense its symbolism and feel a new breath of freedom from our current cultural claustrophobia and be awakened once again to the mythic dimension of man.[65]

The mythic dimension of man, however, is inseparable from the mythic dimension of Earth for She is the referent, the primary law, and She must be honoured as such.

And then it happened.

Like the Delphi of old, She reclaimed Her prophetic voice.

Part II

From Moon To Earth

I love him who is ashamed when the dice fall in his favour and who then asketh: "Am I a dishonest player?" – for he is willing to succumb.

– F. Nietzsche

And we thought that the Earth
Stretched out
Like pancake

And we thought the salmon
Would breed no-matter-what
In the rough and tumbling
Great Pacific rivers

And we thought that the Sun
Would stir life in the bud
Until prayers
Die out

Carpe Diem
We said
In our youthful ignorance
And we kicked and we screamed
And we gorged on the juice
Of eternal fruits

And we felt good
And we felt great
Alive inspired vibrant creative powerful
Sap rising
In our young bodies
Oozing with confidence
Glowing in the dark

Glowing so much in the dark

Where is the dark?
We said
What happened to the dark?

We said
All is light and colours and brightness
This is exhausting
We said
Give us darkness
We said

We so want to rest
And walk the path of trees
And learn the ways of Bear
As we used to
When snow covered our skins
Our furs
Our claws

Wind singing
Through hollow
Bones
Humming the tune
Of bleeding
Lands

Deer wearing
A crown of thorns

Chapter 7

The Earth from Above

The Emotional Reactions of the Astronauts in Space

The world that existed before Christmas 1968 has passed away as irrevocably as the Earth-centred universe of the Middle Ages.
– Arthur C. Clarke

In 1968 'Les Frères Jacques', a group of four French artists, released a song entitled *The Moon Is Dead*. Anticipating the Moon landing by a year, it was a lament, a poignant reflection on the impending moonwalk, in which Pierrots, poets and black cats were mourning the dead Moon. 'Pierrot', an all-white clownish character of the *commedia dell'arte*, has a strong association with the Moon, as in the famous nursery rhyme *Au Clair de la Lune, mon ami Pierrot*. I first heard *The Moon Is Dead* when I was a child. It is the kind of song that transcends time and can be appreciated and loved by anyone of any age. The Frères Jacques were exquisite artists, mixing music, poetry and theatre. Their interpretation of this song is extremely moving and tender.[1] Much later on, in my twenties, I heard it again, as if for the first time, and I remember how I was struck by it, although I did not know why it would touch me so much. Today I do. I am strongly indebted to the Frères Jacques for inspiring my work. They planted a seed in me, they awakened something deep and fragile and meaningful. They anticipated my own Moon landing, too.

By stepping on the Moon, we broke into the house of myth. The Moon had no doubt long lost her mystical status, but she had retained something of her mystery in popular culture. One does not get rid of thousands of years of worship so easily:

Perhaps the fact that the mentalities of the 'primitive' and 'modern' respond so similarly to the sort of sacredness expressed by the Moon can be explained by the survival even in the most totally rationalist outlook of what has been called 'the nocturnal domain of the mind'. The Moon would be appealing to a layer of man's consciousness that the most corrosive rationalism cannot touch.[2]

Which does invite the question: *When exactly do we step in and out of mythological territory?* In other words, when are we in the presence of a working mythology? This question occupied Joseph Campbell a great deal. By way of answer, Campbell proposed that myths fulfilled four main functions, one of which was cosmological:

Of representing the universe and whole spectacle of nature, both as known to the mind and as beheld by the eye, as an epiphany of such kind that when lightning flashes, or a setting sun ignites the sky, or a deer is seen standing alerted, the exclamation 'Ah!' may be uttered as a recognition of divinity.[3]

While most of this definition can be easily accepted by the modern mind it is the last part of the sentence that is problematic, for 'divinity' has become somewhat of a dirty word. As Dawkins and Cox are all too happy to emphasize,[4] 'Ah!' does not have to come bundled up with God. One may be utterly entranced by the physical universe without needing any metaphysical or spiritual explanation.

The astronauts were acutely feeling this tension between Reason and Revelation. "If I get lost in wonder at the sight of a sunset," said veteran astronaut Wally Schirra, "I waste the flight and maybe my life."[5] In *To Touch the Face of God*, Kendrick Oliver relates the telling story of Scott Carpenter. Carpenter was a Mercury Seven astronaut and he was prone to bouts of enchantment and ecstasy. His Sun/Venus conjunction in Taurus symbolised a heightened sensitivity to beauty. Beauty, as James Hillman was keen to stress, is "inherent and essential to soul". It is the manifest *anima mundi*, the soul of the world: "Beauty is the very sensibility of the cosmos, that it has textures, tones, tastes, that it is attractive."[6] By cosmos Hillman meant *kosmos*, which in its Greek sense carried a notion of aesthetics – as in *cosmetic* – an ordered arrangement, a polyphony. This aesthetic notion, said Hillman, had been lost with the Romans, who turned *kosmos* into *universe*, "a collective, general, abstract whole."[7]

Carpenter described his Mercury flight as "the supreme experience of my life". He was more Greek than Roman:

> The sight was overwhelming. There were cloud formations that any painter would be proud of – rosettes or clustered circles of fair-weather cumulus down below… I could look off for perhaps a thousand miles in any direction, and everywhere I looked the window and the periscope were constantly filled with beauty… Later on, I saw the beginnings of the most fantastically beautiful view I have ever had – my first sunset in space.[8]

His raised spirits were his downfall: too busy marvelling at the view, he misjudged his fuel consumption and had the misfortune of landing his spacecraft 250 miles away from the projected landing site. NASA was not amused, driving Flight Director Christopher Kraft to bluntly declare: "That sonofabitch will never fly for me again!"[9] And he never did. Carpenter's experience is said to have had a strong influence on other astronauts, who realized that they'd better keep a tight rein on their potential ecstasies. Roger Chaffee, who tragically lost his life in the *Apollo 1* fire, summed it up shortly before his death: "Imagination must be held in check by a consideration of what is logical and useful, otherwise it becomes a childish instrument.

And none of us are children."[10]

The responses of the *Apollo* astronauts to seeing the Moon at close quarters were reflecting this tension, some of them downright dismissive while others showed strong emotions. For Armstrong,

> of all the spectacular views we had, the most impressive to me was on the way to the moon… We were still thousands of miles away, but close enough so that the moon almost filled our circular window. It was eclipsing the sun, from our position, and the corona of the sun was visible around the limb of the moon as a gigantic lens-shaped or saucer-shaped light, stretching out to several lunar diameters. It was magnificent, but the moon was even more so. We were in its shadow, so there was no part of it illuminated by the sun. It was illuminated only by earthshine.[11]

On board *Apollo 8*, the first mission to fly to the Moon – staying sixty miles off the surface – the crew were less enamoured with the view:

> It appeared to be a vast and forbidding place – in Borman's words, "a great expanse of nothing that looks rather like clouds and clouds of pumice stone"; in Anders's words, "a dirty beach, grayish-white and churned like sand. It looked much like a battlefield, hole upon hole, crater upon crater… It was completely bashed."[12]

Apollo 11's Michael Collins had a sense of dread: "this cool magnificent sphere hangs there ominously, issuing us no invitation to invade its domain."[13]

But a shift seemed to take place when approaching the surface:

> For those astronauts who were tasked with an actual landing,

the moon retained an aesthetic appeal. Cruising over the lunar surface, they found a quality of grandeur in its austerity: "we could not pull our gaze away from the window," remembered David Scott, commander of Apollo 15. There was, later attested by Apollo 16's Charlie Duke, "a beauty about this wasteland. It was spectacular, and I was nearly overwhelmed emotionally."[14]

Back into the Lunar Module after his moonwalk, Armstrong was lost for words: "I thought about the magnificence of the whole thing, but that's difficult to capture in a single description."[15]

The ineffability of the experience was a major hurdle for the astronauts, who were struggling to find the words to convey their emotions. "We weren't trained to smell the roses," Buzz Aldrin explained in his characteristic no-nonsense prose,

> or to utter life-changing aphorisms. Emoting or spontaneously offering profundities were not part of my psychological makeup anyhow. That's why for years I have wanted NASA to fly a poet, a singer, or a journalist into space – someone who could share the emotions of the experience and share them with the world.[16]

Michael Collins thought that "a philosopher, a priest, and a poet" would be the ideal *Apollo* crew although, in a very down-to-earth way (sic), he tempered: "Unfortunately they would kill themselves trying to fly the spacecraft."[17]

I find the astronauts rather severe with themselves: in my opinion, they did a decent job of emotionally conveying their extraordinary travels. Moreover, they had to contend with our own expectations, back on Earth. Here was this unique breed of human beings, who had left Earth and had come back to tell the tale. A hell of a tale it had to be! Nothing short of the highest hyperbole was going to satisfy us. *Apollo 12*'s Alan Bean thought that people projected too much into spaceflight: "I have to admit that if I ran into somebody who had climbed the Matterhorn, I would probably say, 'What is it like?' The answer might reasonably be, 'It is just like climbing all the other difficult mountains I've ever climbed.'"[18]

I hear what Bean says but I am not sure that I agree with him. While no *bona fide* mountaineer myself I do enjoy hiking and I am not too afraid of heights, so under proper guidance I have over the years climbed a number of peaks in the Alps. One of them was the Matterhorn, the crowning achievement of my modest alpinist career. One thing that I would certainly not say about the Matterhorn is that it is like all other difficult mountains. Each mountain comes with its own personality and each climb is different. The Matterhorn is unique in its majesty, towering above the beautiful village of Zermatt, a pyramid rising from the ground in a 1,000-metre impulse, a giant flame of ice and rocks darting towards the heavens. It is an intimidating sight, an awe-inspiring display of *Sereti*, rendered all the more portentous by the tragic history of its first climb. In 1865, the Matterhorn was still resisting any climb, the last great peak of the Alps yet to be "vanquished",[19] a mountaineering conundrum that many attempts had not been able to resolve. Eventually Englishman Whymper, alongside a roped party of six companions, found the key, reaching the summit on July 14 of that year. The rest of the story belongs to mountaineering lore. The conquerors stayed at the top for one hour and proceeded to descend. The descent is often the riskiest part of any climb, but on the Matterhorn, it is particularly treacherous. For safety reasons, the least experienced climbers always descend first, secured by those above them. On that day Englishman Hadow – amazingly for such a climb, a rookie alpinist! – proceeded first. Whimper then recounts the tragedy:

From the movements of their shoulders it is my belief that

[French guide] Croz was in the act of turning round to go down a step or two; at this moment Mr. Hadow slipped, fell on him, and knocked him over. I heard one startled exclamation from Croz, then saw him and Mr. Hadow flying downwards; in another moment Hudson was dragged from his steps and Lord F. Douglas immediately after him. All this was the work of a moment; but immediately we heard Croz's exclamation, Taugwalder and myself planted ourselves as firmly as the rocks would permit; the rope was tight between us, and the shock came on us as on one man. We held; but the rope broke midway between Taugwalder and Lord F. Douglas. For two or three seconds we saw our unfortunate companions sliding downwards on their backs, and spreading out their hands endeavouring to save themselves; they then disappeared one by one and fell from precipice to precipice on to the Matterhorn glacier below, a distance of nearly 4,000 feet in height. From the moment the rope broke it was impossible to help them.[20]

Later on, that day, the three surviving climbers, while searching for their missing companions, saw a giant arch and two enormous crosses appear in the sky. The supernatural seemed to have gatecrashed the party. Four climbers fell to their death on this first ascent and the rumour mill, as often in mountain catastrophes, went into overdrive. Whymper and the Taugwalders, father and son, were accused of deliberately cutting the rope to save their bacon. Traumatised by the experience, wounded by the accusation, Whymper never set foot on a mountain again. Back in England, the public eagerly followed the news of the tragedy. More than any other climb, the Matterhorn's first ascent shaped the legend of the Man-eating Mountain and placed alpinists in the category of reckless daredevils. "Why do you climb mountains?" someone once asked George Mallory, who eventually lost his life on the Everest in 1924. "Because it's there," was the reply.

That reply may be just as relevant for the Moon landing. Do astronauts and alpinists belong in the same category, then, as Bean's remark suggests? The new conquistadores? For what it's worth I personally find a major difference. The space battle between the USA and the USSR was a proxy international war. Supremacy in space was a matter of national prestige and a blatant demonstration of superiority. After World War II, while the American and Soviet big boys were turning their eyes to the heavens, Europe was licking its deep wounds. Germany, who had been at the forefront of rocket science through the ominous V2 program, had lost her best rocket engineers to the USA – chiefly among them Werhner von Braun, who was instrumental in building the giant *Saturn V* rocket that set *Apollo 11* on its course. Desperate to bolster national pride, the European nations had to settle for lesser heights. The Himalayan giants beckoned, and another proxy war took place. The main European powers carved out the Himalayas between them – the Yalta of mountaineering – ensuring that each nation had its consolation prize: the French were the first to go over 8,000m (Annapurna 1950, 8,091m), the English claimed the highest peak (Everest 1953, 8,848m), the Germans the most lethal (Nanga Parbat 1953, a.k.a. the *Killer Mountain*, 8,125m), the Italians the second highest and most technically difficult (K2 1954, 8,611m). The best alpinists of their generation were involved: Gaston Rebuffat, the exquisite French climber, Hermann Buhl, the Austrian trailblazer, Walter Bonatti, the sensitive Italian. You have to take your hat off to the British, though: Edmund Hillary may not have been among the best climbers of his time but his joint ascent with the Nepalese Sherpa Tenzing Norgay was a powerful symbol. It respectfully acknowledged the role played by the local guides and porters in these successful climbs. For once, a Western power was willing to share the limelight. Each and every one of these first ascensions is packed full of drama. While feats of physical endurance had to be performed in these extreme environments, it is the exacerbated relations within members of the expeditions that make for fascinating reading. Behind the official accounts of heroism and male camaraderie lurked jealousy, betrayal and pettiness. It is as if the alpinists never managed to go beyond the strict role of national flag-bearers, at best, or of thrill-seekers with giant egos, at worst. By contrast the astronauts, by leaving Earth, may be said to acquire a new dimension, almost by resonance: they don the mantle of representatives of the whole human race. Despite their attachment to their respective countries, they seemingly rise above national boundaries, performing a service for humankind in its totality. On the first visit to the Moon by *Apollo 11*, a plaque was left on the surface "for eons to come," said Aldrin:

> Depicting the two hemispheres of the Earth and dated July 1969, the plaque stated our heartfelt desire:

HERE MEN FROM THE PLANET EARTH FIRST SET FOOT
UPON THE MOON.
WE CAME IN PEACE FOR ALL MANKIND.[21]

It may be tempting to be cynical about this statement of intent given the geopolitical context underlying the space race, but I do believe that Buzz was genuinely expressing his 'heartfelt desire' when he unveiled the plaque. I, for one, never heard any of the Himalayan conquerors say that they had climbed 'for all mankind'.

The Matterhorn ascent itself is not technically difficult[22] but the length, altitude and sense of foreboding make it a serious endeavour. One starts in pitch darkness from the Hörnli refuge at the base, to then proceed via the north-east ridge. At first the only visible lights are the little dances of the mountaineers' headlamps, above and below. The greatest danger of the mountain is its instability. The rocks are quite loose, often likened to a pile of plates on top of one another, and alpinists dislodging rocks on to the lower climbers cause most accidents. As one rises, with the ground receding further and further away in the distance, the air thins out and the cold bites harder. After a couple of hours, the rising sun gently illuminates the slopes, a soft light that brings the mountain to life. "You great star!" enthused Nietzsche's Zarathustra, "what would your happiness be, had you not those for whom you shine?" It usually takes four to five hours to reach the summit, a razor-thin vantage point overlooking 4,000 feet of void spreading in all directions. This is literally breathtaking, a staggering impression of being suspended in the sky. It took my guide and myself seven hours to climb down. Unusually, the whole mountain was plastered with snow, so much so that the Zermatt guides, out of caution, had decided not to go up that day. My guide was from Chamonix and did not have such reservations but paradoxically, by emptying the mountain of many climbers and thus reducing the risk of falling rocks, the snow made it much safer. But, having to use crampons for most of the climb, it also severely slowed down the pace of ascent and descent. The round trip took us thirteen hours, with very little rest. I was exhausted but radiant. It was my third attempt on the mountain and I had at last made it to the summit – what's more I had been in top form all along and thoroughly enjoyed the climb. It had been a glorious day.

The impression that lingers after a successful climb is quite unique. One often talks of peak experience and of the depression that follows – or maybe deflation is a better word. I do not feel that way after hiking, quite the opposite. An incredible feeling of peace sets in, a profound contentment, the deep sensation of being alive with every fibre of my being, of being at rest in the world. This is why *I* climb mountains. And I think this is why most alpinists do. It is like momentarily putting the demons to rest. This elated feeling, of course, does not last very long. It seeps through the cracks of day-to-day life, inexorably, like water through a broken jug. And so, on to the next climb to refill the jug! A proper Sisyphean task.

Buzz Aldrin knows a thing or two about peak experiences and the difficult times that follow: "I travelled to the Moon," he later confessed, "but the most significant voyage of my life began when I returned."[23] *Apollo 17*'s Gene Cernan felt that by going to the Moon, astronauts had "broken the familiar matrix of life and couldn't repair it." They were at times "unable to focus on minor problems back on Earth,"[24] he said.

The Moon is always more than the Moon: She is a mediator, a crossroads, a liminal space. Michael Collins acutely felt this sensation of arriving at a juncture:

As we turned, the Earth and the Moon alternatively appeared in our windows. We had our choice. We could look toward the Moon, toward Mars, toward our future in space, toward the New Indies, or we could look back toward the Earth our home… We looked both ways. We saw both, and I think that is what our nation must do.[25]

Once on the surface of the Moon, a threshold was reached. Something happened then, something that NASA itself uncharacteristically acknowledged:

Those two words [Aldrin's 'Magnificent Desolation'] summed up the yin-yang of the Moon. The impact craters, the toppled boulders, the layers of moondust – it was utterly alien. Yet Tranquillity Base felt curiously familiar, like home. Later Apollo astronauts had similar feelings. Maybe this comes from staring at the Moon so often from Earth. Or maybe it's because the Moon is a piece of Earth, spun off our young planet billions of years ago. No one knows; it just is.[26]

These few lines, buried deep inside the NASA website, are

remarkable. It feels as if NASA dropped its guard for a few sentences and, away from technical jargon, let its heart speak, bowing to the Mystery. "Yin-yang", "felt", "feelings", "maybe", "No one knows": this is not typical NASA vocabulary. Yet by associating the Moon with yin and yang, NASA was spot on. The Moon carries polarity in her dark and light monthly dance. The Moon carries duality, too, in this instance through her curious mix of utter strangeness and familiarity.

As the NASA account emphasizes, though, as soon as the astronauts set foot on the Moon they felt the presence of Earth. *She started to call back.* And by turning their gaze back to Her, these no-frills engineers heeded that call:

Aldrin: "I could see our shining blue planet poised in the darkness of space."

Armstrong: "It suddenly struck me that that tiny pea, pretty and blue, was the Earth. I put up one thumb and shut one eye, and my thumb blotted out the planet Earth. I didn't feel a giant, I felt very small."

This latter intimate observation by Armstrong is eerily reminiscent of the beautiful words of the mystic Rabbi Nachman of Breslav:

As the hand held before the eye conceals the greatest mountain, so the little earthly life hides from the glance the enormous lights and mysteries of which the world is full. And he who can draw it away from before his eyes as one draws away a hand, behold the great shining of the inner worlds.[27]

While Rabbi Nachman wanted to see beyond earthly life to reach the inner worlds, we can engineer a similar movement and, reaching beyond our thumb, see the Earth for what She is – a home to be treasured.

"Space is a place," said astronaut Charles Walker, "but it is also an all-encompassing experience."[28] Indeed, many astronauts have been profoundly affected by their forays into space and have reported an extraordinary range of experiences. I recently came across a research project at the University of Central Florida under the name of "Space, Science and Spirituality" that proceeded to "investigate, both theoretically and empirically, the effects of outer space travel on the inner space of experiences", with a special emphasis on feelings of "awe, wonder, curiosity and humility."[29] My own curiosity was tickled. It was a worthy effort and it could have been a fascinating study but the dispiriting and dry handling of the subject drove me to absolute despair. 'Awe, wonder, curiosity and humility' became AWCH, a bone-sucking acronym that obliterated the very object of its study. The authors reported 48 categories of experience, later bundled into 34 'consensus categories' – from A, Aesthetic Appreciation, to T, Totality (wholeness of what is experienced). The wording of the whole report was so incredibly technocratic that you have to marvel at the ability of the academic method to suck the juice out of the most sublime experiences: "the hermeneutically-derived categories helped to shape the design of the experiment, the structure of the phenomenological and psychological data collection, and the analysis of the neurophysiological results". Well, maybe. I might be very unfair: there may be value in classifying these experiences with the passion of a forensic entomologist, but I am kind of missing where this debauchery of categories leads into. Once we have dissected the experiences to death, what exactly are we left with?

Frank White would say that we are left with the Overview Effect:

The Overview Effect is a cognitive shift in awareness... It refers to the experience of seeing first-hand the reality that the Earth is in space, a tiny, fragile ball of life, 'hanging in the void', shielded and nourished by a paper-thin atmosphere. The experience often transforms astronauts' perspective on the planet and humanity's place in the universe.[30]

It seems that there are two ways of being in space: floating in orbit around the Earth, which is the experience of most astronauts, or leaving the orbit of the Earth and looking at her from further away, reserved to the *Apollo* pioneers. "Did you find a major difference between your *Gemini* and *Apollo* missions?" Frank White once asked Michael Collins: "There is definitely a different feeling," Collins said:

> At 100 miles up [*Gemini*], you are just skimming the surface, and you don't get a feeling for the Earth as a whole. It's a pity that we have stopped going at a greater distance from the Earth, as with the moon missions. By that, I mean 100,000 miles minimum. When you are in orbit, it's like a roller-coaster ride. On the way to the moon, that feeling of motion stops. It is definitely two very different elements.[31]

Skimming the surface of the Earth, as Collins says, seems to bring a powerful awareness of the unity of life on Earth. But having the bonus of seeing the Earth from afar, in the vastness of space, may induce what White has called the Universal Insight, "an intensification of the Overview Effect that brings a similar understanding of the nature of the universe and our place in it."[32] This is why the 1968 *Apollo 8 Earthrise* picture was so seminal: it brought home a Universal Insight. Although of poor photographic quality, *Earthrise* was the pioneer, our first glimpse, and it shook us to the core. Four years later, in 1972, *Apollo 17* took the magnificent *Blue Marble* picture,[33] showing Earth in her fullness, a pulsing white and blue ball of life, and more than any other picture it was a perfect illustration of the Overview Effect. It duly became an icon for the environmental movement. And then in 1990, *Voyager 1*, which was travelling at 40,000 miles per hour away from the Earth, turned back its cameras towards the receding planets of the solar system and beamed back a picture taken at the staggering distance of 3.7 billion miles.[34] It took 5½ hours for the picture to reach Earth. What it showed was a vast expanse of space, with what looked like beams of light – in actual fact the reflection of sunlight off the spacecraft – among which were dotted very tiny dots, some visible, some not: the planets of the solar system. "This is how the planets would look to an alien spaceship approaching the Solar System after a long interstellar voyage,"[35] said Carl Sagan. Earth was hardly visible at all, a teeny-tiny forlorn bluish spot unremarkably hanging in a beam of light. This picture has been dubbed *Pale Blue Dot* and has inspired Carl Sagan to write the following lines:

> Look again at that dot. That's here. That's home. That's us… Think of the rivers of blood spilled by all those generals and emperors so that… they could become the momentary masters of a fraction of this dot… Our posturings, our imagined self-importance, the delusion that we have some privileged position in the Universe, are challenged by this point of pale light… In our obscurity, in all this vastness, there is no hint that help will come from elsewhere to save us from ourselves.[36]

Sagan's words, for all their sensitivity, carry a strong subtext. In between the lines one can recognise the contours of the contemporary view of the human condition: we are alone, we are insignificant, we are our own masters and our own slaves. Lost in the immensity of space this dot – that's us – is meaningless. In the grand scheme of things, Earth is just too small to have any relevance. The *Pale Blue Dot* picture seemed to confirm our worst fears: "The eternal silence of these infinite spaces terrifies me," Pascal wrote in the 17th century. If *Voyager 1* was to take the same picture today, 30 years later and billions of miles further away, there might be no Earth to see at all – then, what? Which conclusions would we then draw?

By introducing a distance between us and the planet we inhabit, the *Apollo* and *Voyager* pictures have altered our relation to Earth. But it is an alterity in degree. When our eyes stay at ground level, we look no further than our village. In orbit, the view expands to a vast portion of the Earth. The world below looks fragile, a jewel to be treasured: "the minute I saw the view for the first time was one of the most memorable moments of my entire life," said Prince Sultan of Saudi Arabia, who in 1985 flew on board the Space Shuttle *Discovery* as a payload specialist. "I think it has changed my insight into life. I've got more appreciation for the world we live in."[37] With the *Blue Marble* picture, we see Earth in her totality and all of a sudden, we have a vivid symbol of the connectedness of all life on the planet. Further away, with the *Earthrise* picture from the Moon, we sense the connection between Earth and the Universe.

But then, from the confines of the solar system the *Pale Blue Dot* picture, by the incredible scale that it reveals, suddenly seems to point to the meaninglessness of it all: we are no more than a pixel on a photo among hundreds of thousands of pixels. Is there a right distance to look at things, then? Too close and we lose the overall perspective, too far and we lose context – and it is context that gives meaning. This takes us right back to Apollo. The Apollonian sees from a distance, said Hillman. But which distance does he see from? Is it a detached perspective that embraces the whole or that demeans it?

Arendt wrote about the Archimedean point, a vantage reference point – not necessarily geographical – that shifts perspective. The Copernican Revolution was such an Archimedean point, and so was the Renaissance. The perspective birthed by a new Archimedean point informs many levels of our human experience. In *The Master and His Emissary*, McGilchrist highlights how, at the onset of the Renaissance, the rediscovery by Giotto of the perspective in painting, which had been lost for 1,000 years, mirrored the advent of the notion of the individual in society:

> Perspective mediates a view of the world from an individual standpoint – one particular place at one particular time, rather than a God's-eye 'view from nowhere'... [Perspective is] understood differently by the two hemispheres. [It] is, on the one hand, the means of relating the individual to the world and enormously enhancing the sense of the individual as standing within the world, where depth includes and even draws in the viewer through the use of the imagination; and, on the other, a means of turning the individual into an observing eye, a geometer coolly detached from his object's space.[38]

The right hemisphere brings relatedness, the left hemisphere detachment. The ability to detach oneself is, however, a double-edged sword:

> the sense of the individual as distinct from the society to which he belongs enables both an understanding of others as individuals with feelings exactly like one's own, the grounds of empathy; and, at the same time, a detachment of the individual from the world around him that leads ominously to the direction of autism.[39]

It is as if, therefore, the *Blue Marble* and *Earthrise* pictures, by highlighting the connection of the individual to Earth and to the Universe, brought about a feeling of empathy and wholeness, while the *Pale Blue Dot*, by severing this connection through its sheer scale, degenerated into autism and alienation. Things do not happen in isolation: does this evolution of our pictures of Earth, from 1968 to 1990, reflect the evolution of our consciousness towards Earth? "I am less optimistic about getting humankind together than I was 10 years ago,"[40] said Michael Collins. Will we be gifted another picture to shift that consciousness once more? Or, rather, will our changing consciousness be symbolised by a new picture?

While for many astronauts their experience in space has matured into a much deeper relation to the Earth, for others, albeit far less numerous, it has brought about a full-blown spiritual awakening. Two of the most oft-quoted such experiences are those of *Apollo 9*'s Rusty Schweickart and *Apollo 14*'s Edgar Mitchell.

Schweickart was in orbit going on an EVA (extra vehicular activity), commonly called a spacewalk, when his crewmate Dave Scott's camera malfunctioned for a few minutes. Standing on the outside of the Lunar Module Schweickart found himself idly suspended in space, waiting for Scott to fix his tool. His mind momentarily freed from any task, looking down at Earth, he suddenly had the strong intuition that he was the 'sensing element for man', a representative of the whole human race, and it induced in him a profound feeling of humility and responsibility:

Do you deserve this fantastic experience? Have you earned this in some way? Are you separated out to be touched by God, to have some special experience that others cannot have? You know that the answer to that is no. There is nothing that you have done that deserves that, that earned that; it's not a special thing for you. You know very well at that moment, and it comes to you so powerfully that you're the sensing element for man... When you come back, there is a difference in the world now. There's a difference in that relationship between you and the planet and you and all those other forms of life on that planet, because you've had that kind of experience.[41]

On his return Schweickart took up Transcendental Meditation. Seen escorting Maharishi Mahesh Yogi on a tour of NASA's headquarters he was subsequently described by a Houston radio host as "the closest thing to a freak astronaut". I would take that as a compliment, and I hope he did. In 1985 he founded the Association of Space Explorers, which seeks "to provide its members with opportunities to communicate their unique perspective of Earth to help stimulate humanity's sense of responsibility for our home planet."[42] On the front page of its website are found the words of *Apollo 15*'s Al Worden:

Now I know why I am here
Not for a closer look at the Moon
But to look back, at our home,
The Earth.

A few years later, Edgar Mitchell's experience while travelling back to Earth after his moonwalk raised the bar further:

What I experienced during the three-day trip home was nothing short of an overwhelming sense of universal connectedness. I actually felt what has been described as an ecstasy of unity... There was the sense... that there was an intelligence process at work. I perceived the universe as in some way conscious... I realized that the story of ourselves as told by science – our cosmology, our religion – was incomplete and likely flawed. I recognized that the Newtonian idea of separate, independent, discreet things in the universe wasn't a fully accurate description. What was needed was a new story of who we are and what we are capable of becoming. Within a few days my feelings about life were thrown in the air and scattered about.[43]

Mitchell initially did not know what to do with this ecstatic realization. His engineering background had not prepared him for the spiritual dimension of space travel. While always taking great pains to coat his experience in scientific garbs, Mitchell eventually felt that it could be likened to 'savikalpa samadhi', described by the great Indian mystic Patanjali as a meditative state in which human consciousness disappears and everything appears at once in its totality.

Mitchell's experience raises the question of mysticism and space travel. Was what Mitchell experienced a proper mystical experience? In *The Varieties of Religious Experience*, written at the very beginning of the 20th century, American pioneer psychologist William James identified four essential qualities of a mystical experience: ineffability, noetic quality (revealing new knowledge), transiency, and passivity.[44] One may say that Mitchell's experience qualified on all counts. But defining mysticism, an experience that in many ways is by its very nature undefinable, presents a genuine conundrum. In the 12th century, Sufi mystic Suhrawardi stressed the ineffability of his spiritual ecstasies: "As it is said, *al-halu la yu'rafu bil-qal* – 'the states cannot be told by words' – it is not possible to express such states so that others can appreciate or even imagine them."[45]

At the same time as the *Apollo* astronauts were unwittingly exploring inner realms through outer travel, another group of explorers was at work, busily charting their inner landscapes with the aid of various mind-altering substances – psychedelics, as they have come to be known. And they, too, were faced with the same ineffability and noetic quality of their experiences. "How could these layers upon layers," said historian of religions Huston Smith of his mescaline experience, "these worlds within worlds, these paradoxes in which I could be both myself and my world and an episode could be both momentary and eternal – how could such things be put into words?"[46]

The nature of the mysticism revealed by psychedelics has been repeatedly challenged. In the late 1950s a debate opposed

Aldous Huxley, who had extolled in *The Doors of Perception* the 'gratuitous graces' offered by mescaline, to Christian scholar RC Zaehner who objected in a paper called "The Menace of Mescaline" that artificial interference with consciousness had nothing to do with the beatific vision of the Christian mystics. Does 'chemical mysticism', to use Alan Watts' words, qualify as genuine mysticism, then? One may just as well wonder if *ascetisme* or sleep deprivation – traditional techniques used by more religious-minded people – are not also a way of artificially inducing non-ordinary states of consciousness. Where do the astronauts' experiences fit in this debate? Frank White held that the core experience of being in space "include[d] changed perceptions of space, changed perceptions of time, silence and weightlessness."[47] Do these very real physical effects mean that Mitchell's experience must be written off as mere delusion – a temporary alteration of brain chemistry? Are psychedelics, intense meditation, and space travel, mind-distorters or do they expand consciousness by breaking apart the constraining ego structure? It may of course be remarked that years of dedicated Zen practice or numerous space travels may yield nothing in terms of mystical experience; while the ingestion of a small blotter is a quick and effective way to skip past the Angel with the flaming sword and force open the gates of Heaven. This might be the strongest objection to the use of entheogens:[48] there is a sense that one has not paid one's dues before being offered 'the pure gifts of the Gods', as Jung used to call them. The numerous acid casualties of the Sixties, however, are testament to the fact that a heavy price has sometimes to be paid for cutting corners: "Ecstasy!" exclaimed Gordon Wasson, who won fame for his work on the kykeon:[49]

> In common parlance ecstasy is fun. But ecstasy is not fun. Your very soul is seized and shaken until it tingles. After all, who will choose to feel undiluted awe? The unknowing vulgar

abuse the word; we must recapture its full and terrifying sense.[50]

One of the defining discoveries of the 20th century has been quantum physics, in which science and spirituality may be said to meet. In 1975, physicist Fritjof Capra explored this encounter in *The Tao of Physics*, a book written after an ecstatic experience on a beach in California – possibly facilitated by the ingestion of some illegal substance, although Capra kept a respectful silence on the subject:

> As I sat on that beach... I 'saw' cascades of energy coming down from outer space, in which particles were created and destroyed in rhythmic pulses; I saw the atoms of the elements and those of my body participating in a cosmic dance of energy; I felt its rhythm and I 'heard' its sound, and at that moment I knew that this was the Dance of Shiva, the Lord of Dancers worshipped by the Hindus.[51]

Capra's description is not far removed from Mitchell's. William Blake of course – and it may be a bit of a cliché – immediately comes to mind: "If the doors of perception were cleansed everything would appear to man as it is, Infinite. For man has closed himself up, till he sees all things thro' narrow chinks of his cavern." Aboard *Apollo 14* Mitchell, it seems, had been momentarily led out of Plato's cavern. Upon his return he founded the Institute of Noetic Sciences, "a non-profit organization dedicated to supporting individual and collective transformation through consciousness research, transformative learning, and engaging a global community in the realization of our human potential."[52] Worthy ideals, all, but as usual the devil is in the details. A few years ago, I attended a conference in London in which the Head of the Noetic Institute, a woman in her early forties, was speaking. I was full of glorious expectations.

Alas! "The realization of our human potential," she proceeded to explain, had all to do with machines. Her utmost faith in technology was heartbreaking. Even compassion, she asserted, could be programmed into artificial intelligence. As if we as a species had any idea of what compassion was! The whole thing felt eerily close to the transhumanist agenda, and I left bitterly disappointed. I wondered if Mitchell endorsed this view, or whether the Institute had changed course over the years.

Huston Smith once challenged his students to guess, between two accounts of religious experience, which one was drug-induced and which one was not:

1. Suddenly I burst into a vast, new, indescribably wonderful universe… The knowledge that has infused and affected every aspect of my life came instantaneously and with such complete force of certainty that it was impossible, then or since, to doubt its validity.
2. All at once, without warning of any kind, I found myself wrapped in a flame-coloured cloud… Directly afterwards there came upon me a sense of exultation, of immense joyousness accompanied or immediately followed by an intellectual illumination impossible to describe.[53]

And, said Smith, "twice as many students answered incorrectly as correctly."[54] I did too.[55]

Once the dust has settled on the thorny issue of what constitutes mysticism or not, we may say that mystics, poets, psychonauts and astronauts are meeting in the most extraordinary place.

A place that speaks of Eternity in a grain of dust, and where monsters, horses with wings, dwarves, Angels, Unicorns, spirits, dragons and crumbling towers have a story to tell.

A place where one follows with awe the lead back to the Original Seed, while the fabric of the Universe dances in the eyes of a caterpillar.

Ah, not to be cut off,
not through the slightest partition
shut out from the law of the stars.
The inner – what is it?
if not the intensified sky,
hurled through with birds and deep
with the winds of homecoming.
– Rainer Maria Rilke

Chapter 8

Hekate

The Transforming Power of the Moon

It might be said that the Moon shows man his true human condition; that in a sense man looks at himself, and finds himself anew in the life of the Moon.
– Mircea Eliade

In the Borgia Chambers of the Vatican hangs a painting, *Isis with Hermes Trismegistus and Moses*, by Renaissance artist Pinturicchio. The artist has represented the goddess Isis teaching to the mythical founder of the esoteric arts, Hermes Trismegistus, seated to her right, and to the prophet of the Abrahamic religions, Moses, to her left. As Campbell notes, "the statement implied here is that the two variant traditions are two ways of rendering a great, ageless tradition, both issuing from the mouth and the body of the Goddess."[1] This is a remarkable painting, all the more so for its perplexing location – the Vatican, of all places. Here is an astonishing recognition of the ancient feminine wisdom imparted to men – variously known as Sophia, or Shekinah, and in more recent times pictured as the High Priestess card of the Tarot deck, Arcana nr 2. This precedence of the Feminine mirrors the precedence of the Moon. "In Egypt, as in Babylon," said Esther Harding,

> The worship of the moon preceded that of the sun. Osiris, god of the moon, and Isis, moon goddess, sister and spouse of Osiris, and mother of the young moon, Horus, appear in the religious writings prior to the Fifth Dynasty (c. 3000 BC) while the worship of Ra, the sun god, was not established

until late in the Twelfth Dynasty, probably around 1800 BC.[2]

True to her fluid nature, the Moon has been many things to many people, and her place in culture and religion has greatly fluctuated – a vast, cosmic ebb-and-flow. Reflecting the Moon herself, I am tempted to trace the contours of a giant lunar cycle: born at the dawn of humanity – a New Moon – the Moon as divinity reached the apex of her waxing cycle – the Full Moon – during the late Neolithic, at which point her worship was associated with the agricultural cycles of the Earth. She then entered her waning phase, gradually giving way to Sun worship, until a famous day in July 1969 when she was plunged in utter darkness by her symbolic death to the Sun God Apollo – the Dark Moon phase. I may be accused of greatly simplifying and ignoring the incredibly varied cultural and spiritual developments of humanity that took place in so many different places, and I accept the criticism. But I cannot resist the poetry of such an aesthetic symbolic line and I am anyway not pretending to be a historian. It would suffice me to be a symbolist, and one with a poetic licence at that. Which is a convenient justification for saying anything you like and getting away with it. Yet I like to believe that there is some logic to the madness.

The Moon was often pictured in a triple aspect, with different goddesses associated with each aspect. This triple figure symbolised the three main stages in a woman's life: Persephone/ Artemis for the New Moon/Crescent Moon (the Maiden), Demeter/Hera/Aphrodite for the Full Moon (the Mother), Hekate for the Waning/Dark Moon (the Crone). I want to spend some time with the latter, whom I referred to in Chapter 3 as the lunar Feminine in contrast to Athena, the solar Feminine.

As the incarnation of the dark phase of the Moon, Hekate was the goddess of "magic, witchcraft, the night, moon, ghosts and necromancy."[3] For the ancients, says Harding, the dark of the moon was a very dangerous time: "the destructive powers were at their height. Floods and storms, or destructive pests, were to be expected, and ghosts walked or flitted shrilling through the air."[4] Hekate was often contrasted with Artemis:

> Hekate and Artemis were two intrinsically separate goddesses who often came together as a dual Moon goddess: Artemis, young, wild and beautiful, standing at the beginning of the cycle, and Hekate, older, prophetic and deathly, standing at the end.[5]

Hekate was well known in myth for having assisted Demeter in her quest for her daughter Persephone, abducted by Hades. She was subsequently often represented carrying torches – the torches that she had used to guide Demeter through the night. After the recovery of Persephone, Hekate remained as her companion and in this function became a deity of the lower realm. Much like Hermes, she could travel the worlds. As Cashford remarks, "in the lunar cycle the dark is not only of death, so it is no surprise that Hekate's torches were carried around the fields to make them fertile, and that she was invoked as midwife and nurse at childbirth, bringing visions as well as lunacy and death."[6] She was a feared goddess and had to be properly propitiated. As with so many deities, however, Hekate's image markedly changed over the centuries. In Christian times, says Cashford,

> as the Virgin Mary grew in the hearts and minds of Christians, [Mary] assumed the shining mantle of the heavenly Moon, and with it much of the ancient imagery of the lunar goddesses who had gone before her: Isis, Artemis, Persephone and Aphrodite, among others. Hekate, the Dark Moon, was officially given to witchcraft and the devil.[7]

In Ancient Greece, one of her main attributes was as the goddess of crossroads. She occupied liminal spaces, and in this function was further associated with Hermes:

Hekate traditionally was supplicated at the earthly crossroads to insure safe transition through an uncertain point; she was the factor that bridged the gap imagined to exist there, guiding men through a place that was proverbial for its uncertainty.[8]

By reaching the Moon we have come face to face with Hekate, not only in her role as goddess of the Dark but also as goddess of the Crossroads. For the Moon acts as a liminal space, a frontier, a role that she has occupied since Plato and Aristotle divided the cosmos into two regions: that above the Moon, eternal and unchanging, and that of the sub-lunar world, from Moon down to Earth, the imperfect realm of changes and generation – a world of suffering, of *pathos*, as Eliade calls it:

The sub-lunar world is not only the world of change but also the world of suffering and of 'history'. Nothing that happens in this world under the moon can be 'eternal', for its law is the law of becoming, and no change is final; every change is merely part of a cyclic pattern.[9]

In the classical tradition, the Moon is seen as a stage on the spiritual journey to the Sun, the ultimate Good. "The idea of the Moon as a door, gate or mirror, opening on to 'the next' or a further world, is widespread, and then the Moon becomes a dividing line between the temporal world of becoming and the eternal world of being, serving as a place of transition between the two."[10] For Plutarch, the relation between Earth, Moon and Sun was likened to an organism in which the Moon acted as an intermediary principle:

The Sun acts as a heart, and sheds and distributes out of himself heat and light, as it were blood and breath. Earth and sea are to the Universe, according to Nature, what stomach and bladder are to the animal. The Moon, lying between Sun and Earth, as the liver or some other soft organ between heart and stomach, distributes her gentle warmth from above, while she returns to us, digested, purified, and refined in her own sphere, the exhalations of Earth.[11]

Much later on, in the 20th century, the distributing role of the Moon – receiving the light of the Sun at the New Moon and dispensing it gradually through her monthly cycle – was emphasized by humanist astrologer Dane Rudhyar:

The Moon is a means to an end. She is the mediatrix, mother or Muse... She distributes solar potential (i.e. spiritual food and energy) through organic and psychological agencies which she builds to fit the need of the evolving material units, be they cells or personalities. She is therefore the servant of both earth and sun. She releases the *light* of the sun and by so doing serves the need of earth creatures for organic and psychic life.[12]

This echoes the words of Jesuit theologian Hugo Rahner: "Selene becomes that heavenly star which hangs as an intermediary between the sublime light of Helios and the dark earth, the great mediator between the world of pure spirit of the fixed stars and the dark sensuality of the earthly elements."[13] At the time of the dark of the Moon, Sun and Moon are conjunct. Selene receives Helios in sacred union. It is then "as though Selene became pregnant with the light of Helios, and having thus been made fruitful by the sun, she becomes the birth-giving mother of all living things."[14] She mediates the light that She receives, softens it by mixing the fire of the sun with "the water of her own being,"[15] then distributes it.

Dante's cosmology in *The Divine Comedy* is a direct heir to the

Greek classical tradition – as is the whole medieval geocentric cosmos. In the centre was the fixed Earth around which nine concentric spheres revolved. Immediately beyond Earth were the spheres of air and fire, the latter one reaching to the Moon. Beyond the Moon were the planets in ascending order: Mercury, Venus, Sun, Mars, Jupiter and Saturn. Beyond the nine spheres was the *Primum Mobile* – the First Movement – the realm of the fixed stars and an ordering principle that governed the whole. To this physical organisation corresponded a spiritual one. Spirit incarnated in a cascading descent through the different spheres and the aim of the spiritual life on Earth was to re-ascend along the same path in reverse order. At the opening of Canto II of the *Paradiso*, Dante, who is about to take off from Earthly Paradise to ascend to the Moon, addresses the reader as such:

> O ye who is in little bark, eager to listen, have followed behind my ship that singing makes her way, turn back to see your shores again; do not put forth on the deep, for, perhaps, losing me, you would be left bewildered. The waters I take were never sailed before. Minerva breathes, Apollo guides me, and the nine Muses show me the Bears.[16]

Just like Neil Armstrong, then, Dante is taken to the Moon by Apollo. With *Apollo 11* we have physically emulated Dante's spiritual voyage. We have reached the first sphere of the medieval cosmos, the liminal space between being and becoming, and we have come to a crossroads. What does this mean? Which choice are we facing?

We may find the beginning of an answer if we acknowledge the association between Moon and Soul. Earth, Moon and Sun were respectively likened to Body, Soul and Spirit. In this trinity, Soul was intermediate between mind (in the sense of higher mind, spirit) and body: "Soul, a compound and a middle term," said Plutarch, "has, like the Moon, been formed by the god, a blend and mixture of things above and things below."[17] Receiving minds from the Sun, the Moon's function was to fashion new souls, for which the Earth would provide a body. At death the Moon was receiving the departing souls and her distributing function was again highlighted by Plutarch:

> Earth gives nothing after death of what she received for birth; the Sun receives nothing, save that he receives back the mind which he gives, but the Moon both receives and gives and compounds and distributes in diverse functions.[18]

For Plutarch, Man was fashioned through a strictly hierarchical process: mind precedes soul, and soul precedes body. The union of soul with body made up the emotional component, while the union of soul and mind produced reason. Death was a two-pronged process. The first death made man two out of three – the earthly death, in which "Demeter parts soul from body quickly and with force"[19] – the other one made one out of two – the lunar death, the domain of Persephone, "who parts mind from soul gently and very slowly."[20] Before the second process could take place, however, souls wandered between Earth and Moon, and the duration of this wandering depended upon the righteousness of the soul.

The view of the Moon as the abode of the souls was not circumscribed to Rome. As Cashford stresses, it was found all over the world, from Southern Africa to Japan via Egypt and Polynesia.[21] What was maybe more specific to the Western tradition, although not by any means exclusive, was the gradual demotion of the Moon and her increasing association with Hell realms:

> The Moon does not only embody the same circle of death and birth found on Earth (a 'vicious' circle to the solar view); it incarnates more profoundly all the vices of the 'sublunary

realm', the realm of decay and dissolution. Sometimes the Christian 'Hell' was placed on the Moon, or else those sins, for which Christians believed they went to 'Hell', went to the Moon instead.[22]

As highlighted in Chapter 4, the Moon suffered greatly from the Copernican Revolution in that she became subservient to the Earth, herself not in a central position anymore. Her inferiority was glaringly obvious: she became, in Cashford's words, "Earth's wasteland". For Renaissance philosopher Pico della Mirandola, the Moon "is the lowest earth and the most ignoble of all the stars."[23] In the crescent of the New Moon, we were not marvelling at the thin strip of light any longer but became morbidly obsessed with what the darkness concealed: "the loveliness of the new moon," said Jung, "hymned by the poets and the Church Fathers, veils her dark side, which, however, could not remain hidden from the fact-finding eye of the empiricist."[24]

Somewhere along the way, in parallel with the demonization of the Moon, we lost Soul, the intermediary principle. "The Moon is the mediator between the Sun and the Earth," said Rudhyar. "She is, in modern psychological terminology, the *anima* which serves as a link between man's conscious ego and the all-encompassing wholeness of spirit, the God-within."[25] The *anima*, the Latin word for soul, has indeed been a major casualty of the second millennium.

This loss is in sharp contrast to the late classical period and the teachings of the Neoplatonists. Scholar Sarah Iles Johnston has highlighted the increasing association between Hekate and the Moon that took place from the 1st century CE. Under the influence of the Chaldean Oracles (2nd century CE) Hekate's role expanded even further: she became associated with the Platonic Cosmic Soul – the *Anima Mundi*:

Hekate is best known to classicists and historians of religion as the horrific patroness of witches. But from the Hellenistic age onwards, some Greek and Roman philosophers and magicians portrayed her quite differently, allotting to her such duties as ensouling the cosmos and the individual men within it, forming the connective boundary between the divine and human worlds, and facilitating such communication between man and god as could lead eventually to the individual's soul release.[26]

In the *Timaeus* Plato had posited the existence of a Cosmic Soul that infused everything in the universe:

> The soul, which was interwoven throughout the entire fabric from the centre to the furthest limits of the universe, and coated the outside too, entered as a deity upon a never-ending life of intelligent activity, spinning within itself for all time. The soul is invisible (as opposed to the body of the universe, which is visible), and since it is characterized by reasoning and harmony, it is the supreme creation of the supreme intelligible and eternally existing being.[27]

Equated with Soul by the Chaldeans, Hekate, says Iles Johnston, played three cosmological roles:

1. transmitter of the Ideas and therefore structurer of the physical world
2. dividing bond between Intelligible and Sensible Worlds
3. source of individual sources and enlivener of the physical world and of man.[28]

This was a major function, not to be undertaken lightly, and only a deity of terrifying power could handle the charge. But one thing that Hekate was not short of was *Sereti*. She commanded respect and no doubt was worthy of the honour. In her role as cosmic intermediary, said Iles Johnston, Hekate was conveying divine will: "Hekate and Apollo were the two divinities usually credited with the divine messages."[29] How lucky we are: by going to the Moon we have had the privilege of standing in the presence of the two most appropriate divinities to hear the will of the gods – whatever names, attributes or genders we want to give them. But to hear properly we must of course remove the wax from our ears.

As a mediator between the Intelligible and Sensible worlds, Hekate ascended to the daemonic throne. She became the undisputed queen of the *daemones*, these quintessential Greek figures whose importance grew enormously in the first centuries CE. When the divinities retreated further and further away in the Heavens it became quite natural for men to turn to tutelary figures that could intercede on their behalf. The *daemones* were such figures and they became a staple of Neoplatonist thought – in particular with Iamblichus, who developed a whole hierarchy of mediating beings inhabiting the regions between Man and God:

> These classes of beings bring to completion as intermediaries the common bond that connects gods with souls, and causes their linkage to be indissoluble. They bind together a single continuity from top to bottom, and render the communion of all things indivisible. They constitute the best possible blending and proportionate mixture for everything, contriving in pretty well equal measure a progression from the superior to the lesser, and a re-ascent from the inferior to the prior.[30]

For Iamblichus, these intermediary beings performed an essential role in magic and rituals, and in theurgy (literally: the work of God). One had to address them in the proper way so that they may light the path to the Divine – and, more prosaically, so that they would not cause harm. For Iles Johnston the *daemones* were commonly associated with the Moon.[31] With Christianity they became demons, malefic beings, but their function was also taken over by saints and angels, interceding with God on behalf of humans or delivering divine messages. Angels and demons may be two sides of the same coin, as the German poet Rilke recognised:

If my devils are to leave me, I am afraid my angels will take flight as well.

Hekate, Moon, Soul may be seen as different expressions of one particular aspect of human experience – an esoteric correspondence – reaching down to the same root. Yet something else has to be added. As Iles Johnston remarks, "the [Chaldean] Oracle fragments identically describe Hekate and Soul as possessing wombs":

Fragment 96 says:
Soul, being a brilliant fire by the power of the Father,
Remains immortal and is the Mistress of Life
And holds the plenitude of the full womb of the cosmos
Fragments 51 and 52.1:
For all around the hollows of the cartilage of [Hekate's] right flank,
The abundant liquid of the Primal Soul gushes unceasingly,
Completely ensouling the light, the fire, the aether and the Cosmoi.[32]

The full womb of the cosmos: Hekate, Moon and Soul are archetypally feminine. In the vast *solarisation* process of the last four thousand years, it may come as no surprise, then, that their fortunes have followed a similar course. While Hekate, along with the whole Greek pantheon, became relegated long ago to the folkloric past of humanity, the Moon retained her clout for much longer. It took Copernicus and Galileo to kick her off her pedestal for good. In the process she became demonized, acquiring the dark, ominous qualities once attributed to Hekate. Soul, for her part, caught between the rock of the body and the hard place of spirit, was dismissed as an intermediary principle and as an organ of cognition in her own right – something that Jung, Corbin and Hillman spent their lifetimes trying to redress.

I, too, had lost sight of soul. My second night in custody had been the hardest. I was battered, exhausted, humiliated. I had been transferred to the notorious Conciergerie jail in the middle of Paris. The transfer had been done in a police car with sirens blazing. I could see passers-by throwing quizzical looks inside the car – or was it my imagination playing tricks? I felt as if I had lost all grip on my life. All I could do was look at myself from the outside, trying to gain a higher perspective and keep a modicum of humour. I was watching this man in the rear seat of the police car, taken to custody like a criminal, stirring curiosity from the people walking by, and I felt like I was in a very bad B movie, with me in the starring role. In my cell it was freezing cold. The month was October and no blanket had been provided. My feedback would be scathing on TripAdvisor! I was wrapped up in my grey office suit. The cell was grey. I was grey. I did not know it at the time but I was in the presence of Hekate, as goddess of the crossroads, as goddess of the dark, and She was talking. I did not hear her clearly until the following day, but without a doubt, She had come with a message. That message was the message of the soul. Hekate had dispatched my *daemon*, who had had a bit of fun with me and had roughed me up so that I may *listen*. During that night Pluto, Lord of the Underworld, was opposing the Moon in the sky, and their opposition was on my Ascendant/Descendant axis. It did feel like a death of some sort. The Moon was in Gemini, the sign of communication.

Hekate was talking indeed.

One haunting question still lingers though: what exactly do we mean when we speak about 'Soul'? For Hillman,

> Soul, psyche, anima, animus have etymological associations with bodily experiences and are concrete, sensuous, and emotional, like life itself. We have difficulty with these terms because they are not true concepts, rather they are symbols which evoke meanings beyond any significations we give them through definitions.[33]

It is not easy to find one's way between these different notions of the soul. On the one hand is *soul* as in the Christian individual soul (which, of course, would need further qualification). Then comes the *anima* ('soul' in Latin): Jung, who spent a lot of time grappling with this notion, was careful to distinguish the empirical concept of the anima from the Christian idea of the soul – or from any religious notion of soul for that matter. Jung's notions of anima and animus were psychological and highly individual. Hillman was famously unhappy with Jung's definition of the anima as "the contrasexual side of man", which he thought was conceived in a "fantasy of opposites". Hillman was assigning a much wider definition to the anima, beyond gender. Hillman's own psychological practice strived for 'soul-making', which "[implied] the stirring of an emotional and living factor of overwhelming importance for my well-being now and for my death."[34] Where then does psyche ('soul' in Greek) stand in relation to soul and anima? I understand psyche as the wider personality, the sum total of conscious and unconscious elements. Hillman, however, was going further: psyche is not in us, he said, we are in psyche. Which sounds to me very similar to Plato's notion of Cosmic Soul. "Anima becomes the primordial carrier of the psyche, or the archetype of the psyche."[35] Psyche, Anima, Soul, Self: the hall of mirrors of the inner landscapes. Maybe I will settle with Hillman's simple definition of soul as "the inward, downward factor in personality, the factor which gives depth."[36] Depth, yes. We have to *grow down*, Hillman said: the object of depth psychology was soul-making. In my understanding, this inward factor reaches all the way down to the intersection of the personal and the collective – and by collective, I mean transpersonal, too.

In his work Hillman noticed the increasing interlocking of personal and collective pathologies: "I can no longer distinguish clearly between neurosis of self and neurosis of world."[37] Drawing from the Platonic tradition, Hillman wanted to redefine and expand the soul of the world:

Let us imagine the *anima mundi* neither above the world encircling it as a divine and remote emanation of spirit, a world of powers, archetypes, and principles transcendent to things, nor within the material world as its unifying panpsychic life principle. Rather let us imagine the *anima mundi* as that particular soul spark, that seminal image, which offers itself through each thing in its visible form... Not only animals and plants ensouled as in the Romantic vision, but soul given with each thing, God-given things of nature and man-made things of the street.[38]

By thus redefining the *anima mundi*, the world appears in its particulars – shapes, colours, textures, atmospheres. This way of engaging with the world is the way of the right hemisphere, which sees context and pays attention to the individual, in contrast to the propensity of the left to think in abstract and general laws. It creates a new psychic reality that gives back to things their interiority and depth – in short, their soul: "This response ties the individual soul... with the world soul; I am animated by its anima, like an animal. I reenter the Platonic cosmos."[39] Hillman makes a further interesting point: "the appreciation of the *anima mundi* requires adverbs and adjectives that precisely imagine the particular events of the world in particular images, much as the ancient gods were known through their adverbial and adjectival epithets – grey-eyed Athene, red-faced Mars, swift-footed, chaste Artemis."[40] Or cow-eyed Hera.

About twenty years or so ago, I started to write poetry. After producing enough material I went looking for a publisher and found one in the middle of Paris. I was elated, of course, but the path leading to actual publishing was long and tortuous. It took two years of reworking and polishing the text with the publisher, who guided me through the pitfalls of the debutante poet and helped me streamline my words. He was particularly insistent on one point: rid the text of adjectives and adverbs. My natural bent had been very much towards the adjective – Baudelaire had always been a favourite of mine – and I struggled terribly to cut those branches deemed superfluous. The acknowledged master of poets was Rimbaud, and Rimbaud had unequivocally stated: "Il faut tuer l'adjectif."[41] I did my best to comply, and on some occasions, it was worthwhile. But today I wonder. At the risk of finding myself at odds with the French poetic community, I must confess that I have always found Rimbaud's poetry quite cerebral and opaque. When I read Rimbaud, my senses are not tickled. I may be taken aback by the power of the image, by the utter brilliance and *fulgurance*, but I do not feel embodied. With Baudelaire I can smell the rotten carcass by the roadside. This might not be an endearing smell, but it oozes soul as much as stench. "At that moment when each thing," said Hillman,

> each event presents itself again as a psychic reality... then I am held in an enduring conversation with matter. Then grammar breaks its hold: subject and object, personal and impersonal, I and thou, masculine and feminine find new

modes of intermingling... Then Eros descends from being a universal principle, an abstraction of desire, into the actual erotics of sensuous qualities in things: materials, shapes, motions, rhythms.[42]

The organ of cognition of this new psychic reality was the heart – and by heart, Hillman did not mean a 'sentimental subjectivism' or a 'simplistic psychology'. The heart as an organ of cognition had long been superseded by the brain, and in this move we can also perceive the contours of the left/right dichotomy:

When the brain is considered to be the seat of consciousness we search for literal locations, whereas we cannot take the heart with the same physiological literalism. The move to the heart is already a move of *poesis*: metaphorical, psychological.[43]

Long before Hillman, the heart had been recognised by the Sufis as the true organ of perception. Henry Corbin, a French philosopher, theologian, and director of Islamic Studies in Paris in the 20th century, was profoundly influenced by the writings of the Sufi mystics Suhrawardi and Ibn'Arabi in the 12th and 13th centuries. Steeped in Sufi Neoplatonism, which came replete with a whole angelology, Corbin's notion of soul was more religious in orientation but, like Jung and Hillman, he was at pains to re-emphasize her mediatory function. The Western philosophical tradition, Corbin remarked, admitted only two kinds of knowledge, sense perception (body) and intellectual concepts (spirit):

There is our physical sensory world which includes both our earthly world... and the sidereal universe... This is the sensory world, the world of phenomena (*molk*). There is the suprasensory world of the Soul or Angel-Souls, the *Malakut*... [And] there is the universe of pure angelic intelligences

[*Jabarut*]. To these three universes correspond three organs of knowledge: the senses, the imagination, and the intellect, a triad of which corresponds the triad of anthropology: body, soul and spirit.[44]

For Corbin, this intermediate realm, the *Malakut*, was metaphysically necessary: it was the place where "the spiritual takes body and the body becomes spiritualised."[45] This intermediate realm Corbin called the *mundus imaginalis* – the world of the imaginal. A world that could only be known through the *proper* use of the imagination,[46] which then recovered its noetic and cognitive function – a function that had been "left to the poets only."[47] Corbin was fiercely protective of the *mundus imaginalis*, limiting its access to "very specific realms of spiritual and indeed mystical reality."[48] Even Jung and Hillman's active imagination exercises were anathema to Corbin, who refused the profane use of the imagination.

I once worked on an academic essay that set to inquire whether Corbin's *mundus imaginalis* offered an adequate theory for psychedelic visionary experiences. I was curious: did Corbin, who lived through the Sixties, ever consider this question? I contacted "L'Association des Amis de Henry et Stella Corbin"[49] in France and received the following reply: "Thank you for contacting us but I assure you that you will not find anything on this subject in Henry Corbin's work, the word [psychedelia] does not appear in his work and he never commented on the subject."[50] Dead end, then. The email concluded: "I want, however, to emphasize something important. The use of the word *mundus imaginalis* refers to a precise 'metaphysical structure' in the philosophers whom Corbin analyses... If... you use the term in a general sense, you will have to explain yourself." My essay strived to do just that: "Noting that the epistemology of Corbin's metaphysics is at odds with psychedelic visionary experiences," I proposed, "it [is] concluded that both may lead to a similar ontological ground, once one accepts to strip the *mundus imaginalis* of Corbin's precise orthodoxy." "Are the gates to the *mundus imaginalis* too fiercely guarded?" I concluded. "Maybe now is the time for the imaginal world to grant access to more travellers, to all those who, caught between the Charybdis of the senses and the Scylla of the divine intellect, are looking for a way through, an oasis where the *imaginatio vera* may flourish and guide them."

My point here is certainly not to oversimplify. I am conscious that we are entering into very arcane territory: Neoplatonism, Sufism, Mysticism – an intimidating dance of isms. But a few threads are being woven, Moon-like, that produce a rich tapestry: Body-Soul-Spirit; Senses-Imagination-Intellect; Earth-Moon-Sun. The middle realm – the realm of soul and of the imagination – is the realm of the Moon. The outer voyage to the Moon has an inner resonance: when we travel to the Moon, we contact Soul and Imagination. And so, while I have no doubt that Corbin would frown upon my (mis) appropriation of the word, I feel that the outer voyage to the Moon is mirrored by an inner voyage to the *mundus imaginalis*. I would even venture further: we have learnt nothing by going to the Moon if we are not able to look at the whole venture with the incomparable gaze of the Imagination. Collecting rocks is at best knowledge, at worst a distraction. "It is only with the heart that one sees rightly. The essential is invisible to the eye," said Saint-Exupéry's Little Prince. Soul has been awakened. Creation is Epiphany:

We wish to stress on the one hand the notion of the *Imagination* as the *magical* production of an *image*, the very type and model of magical action, or of all action as such, but especially of creative action; and on the other hand, the notion of the image as a body (a *magical* body, a *mental* body), in which are incarnated the thought and will of the soul.[51]

Creation is Epiphany, that is, a passage from the state of occultation or potency to the luminous, manifest, revealed state; as such, it is an act of the divine, primordial Imagination.[52]

Chapter 9

The Sacred Marriage

A Different Perspective on Apollo 11

The factors which come together in the coniunctio are conceived as opposites, either confronting one another in enmity or attracting one another in love.

– CG Jung

A piece of the sacred bread was missing. On the last Sunday before *Apollo 11* was scheduled to take off, a service was held at the Webster Presbyterian Church. The service concluded as usual with the Communion rites: "As [the Reverend] Woodruff broke the loaf of bread and held it for view, he pointed out that the loaf was not whole; he did not say what had happened to the missing piece, but the congregation understood that, symbolically, it had gone with Buzz."[1]

Communion

That little piece of bread made a somewhat remarkable journey: it was taken all the way to the Moon and used by Buzz Aldrin to perform Communion shortly after the *Eagle* had landed. "Weeks before, as the *Apollo* mission drew near," recalled Aldrin in *Magnificent Desolation*, "I had asked Dean Woodruff… to help me to come up with something I could do on the Moon, some symbolic act regarding the universality of seeking."[2] Dismissing patriotic symbols as "trite and jingoistic", Aldrin settled "on a well-known expression of spirituality: celebrating the first Christian Communion on the Moon, much as Christopher Columbus and other explorers had done when they first landed on their 'new world'."[3] One may feel discomfort with the evocation of the

201

conquistadores spreading Christianity on the American shores. I certainly do, and I do not think it has anything to do with any unsophisticated anti-Christian stance on my part. Dark shadows have to be acknowledged. I dread to imagine that the astronauts landed on the Moon with the same mentality, and in fact I do not think they did. For one thing, Armstrong was not interested in the ceremony. He did not approve of it, nor did he condemn it. He just kept busy while Aldrin was proceeding. As for Buzz, I genuinely believe that it never crossed his mind that performing a Christian rite was rather incompatible with laying a stele claiming that the astronauts had come for all humanity. This contradiction also seemed to have escaped the Rev Woodruff: "We dedicate unto Thee, Thy servant and our brother, Edwin Aldrin, to represent the Body of Christ, our nation, and all mankind on the first expedition to another planet."[4] I do not think that there is in these words anything more than a deep sense of entitlement, the unconscious self-evident righteousness that comes from having been the dominant religion in the dominant culture for two thousand years. To be fair to Aldrin, the evident paradox did eventually hit him:

Perhaps, if I had to do it over again, I would not choose to celebrate communion. Although it was a deeply meaningful experience for me, it was a Christian sacrament, and we had come to the Moon in the name of all mankind – be they Christian, Jews, Muslims, animists, agnostics, or atheists. But at the time I could think of no better way to acknowledge the enormity of the Apollo 11 experience than by giving thanks to God.[5]

Deke Slayton, who ran the flight-crew operations for *Apollo 11*, had given his consent to Buzz's ceremony but had made it clear that it had to remain private – and not aired to the whole world as Buzz had intended. NASA had grown nervous about the public expression of faith by its astronauts. A few months before, a civil lawsuit had been filed by atheist Madalyn Murray O'Hair. The reason for Ms O'Hair's wrath was the somewhat overt display of religious fervour by the *Apollo 8* astronauts. Frank Borman, the *Apollo 8* Commander and an Episcopalian lay preacher at his church in League City, Texas, had written a prayer before leaving Earth and had read it in lunar orbit: "Give us, O God, the vision which can see thy love in the world in spite of human failure. Give us the faith to trust thy goodness in spite of our ignorance and weakness. Give us the knowledge that we may continue to pray with understanding hearts." In addition, just as the mission was circumnavigating the Moon on Christmas Eve 1968, the three astronauts had taken turns to read the opening chapter of Genesis. Upon hearing the prayer, an incensed Ms O'Hair had announced that she would take NASA to court for 'evangelizing' by broadcasting prayers from space, and NASA were worried. It was bad PR and an unwanted distraction at a time when the nation was far from united behind the space programme. I wonder what would happen today in similar circumstances. The world has arguably turned more religious since then, and more intolerant too. Social media has obliterated dialogue and transformed it into a mud-slinging match of extreme violence. The Christian right is certainly way more powerful than in the Sixties and in all likelihood would not cave in so easily.

Buzz, however, complied. He started by inviting everyone to reflect on the solemnity of the moment:

So, during those first hours on the Moon, before the planned eating and rest periods, I reached into my personal preference kit and pulled out the communion elements along with a three-by-five card on which I had written the words of Jesus: "I am the vine, you are the branches. Whoever remains in me, and I in him, will bear much fruit; for you can do nothing without me." I poured a thimbleful of wine from a still plastic container into a small chalice, and waited for the wine to settle down as it swirled in the one-sixth Earth gravity of the Moon. My comments to the world would be inclusive: I would like to request a few moments of silence... and to invite each person listening in, wherever and whomever they may be, to pause for a moment and contemplate the events of the past few hours, and to give thanks in his or her own way.[6]

Buzz then turned the microphone off: "I silently read the Bible passage as I partook of the wafer and the wine, and offered a private prayer for the task at hand and the opportunity I had been given."[7] I like the image of the wine slowly swirling around the chalice in the gravity of the Moon. Wine, Chalice, Moon: these are words loaded with symbolic weight. Even 'gravity' lures us into a mysterious realm. For one thing it, too, has weight, as Spike Milligan would no doubt have cheekily remarked. But what is gravity? Nobody actually knows. We may be very good at explaining its *modus operandi*, at predicting its behaviour, especially since Einstein graced the planet with his presence, but we are still nowhere near understanding it philosophically. "Give us one free miracle, and we'll explain the rest" – this, according to Rupert Sheldrake, is modern science's stance, and he has a point. Give us the Big Bang, and we'll take care of the rest. As far as I am concerned, gravity is love. The love of matter for itself, the longing to return to its original point of Oneness – before it was all blown up to smithereens 13 billion years ago. If you lift a stone from the ground, you can almost hear it. The longing! The desperate desire to go back to the ground, to be one with Earth again. Put me down, it says! You can feel it in the palms of your hands, this longing, it weighs you down, it is so pressing and focused and demanding. But it is a symbol too, for we are stones and rocks and we long to be reunited. A powerful meditation of the druids consisted in lying down in darkness with a stone on their belly. Crushed by matter above and under.

Reunited – One again. But I can't shake off the image of the wine gently settling down in the chalice, as if in slow motion. There is real poetry here. Buzz is indeed a poet, it's just that he does not know it! 'Fly a poet to the Moon', he'd said – it's you, Buzz! There's a touching awkwardness in Buzz – like an elephant in a china shop, you feel that you are always on the brink of disaster. On another occasion, Buzz talked of the wine "curling up slowly and gracefully up the side of the cup". There is frailty, vulnerability, an in-built tension in the cracks of which, arguably, miracles can happen. Anyway, I never met the man of course, and I might be all very wrong, but I find him endearing, for some reason. I am grateful to him for bringing back this image of the wine on the Moon, a true offering, like a flower he'd pick up on the lunar surface to enchant us. And the fact that he did not mean it as a poetic act makes it even more poignant. I wish he had filmed the scene though.

The first food and drink consumed on the Moon, then, were, for eternity, consecrated bread and wine. Should we rejoice? It does sound better than a standard meal on board the Lunar Module:

8 bacon squares (IMB=Intermediate Moisture Bite)
Peaches (R=Rehydratable)
6 Sugar Cookie Cubes (DB=Dry Bite)
Coffee (R=Rehydratable)
Pineapple-grapefruit drink (R=Rehydratable).[8]

Bread and wine stand for the body and blood of Christ. Symbolising the Last Supper, they are part of a ritual re-enacting a divine event that took place on Earth. "As to the special nature of these substances," says Jung,

bread is undoubtedly a food. There is a popular saying

that wine 'fortifies', though not in the same sense as food 'sustains'. It stimulates and 'makes glad the heart of man' by virtue of certain volatile substance which has always been called 'spirit'. It is thus, unlike innocuous water, an 'inspiriting' drink, for a spirit or god dwells within it and produces the ecstasy of intoxication. The wine miracle at Cana was the same as the miracle in the temple of Dionysus... Bread therefore represents the physical means of subsistence, and wine the spiritual. The offering up of bread and wine is the offering of both the physical and the spiritual fruits of civilization.[9]

Through the offering of bread and wine, man confirms his status as 'civilized', says Jung: "they represent a definite cultural achievement, which is the fruit of attention, patience, industry, devotion, and laborious toil... Where wheat and wine are cultivated, civilized life prevails. But where agriculture and vine-growing do not exist, there is only the uncivilized life of nomads and hunters."[10] Man's patient work is offered to God. But aside from this cultural and collective dimension, adds Jung, "the union of Host with chalice signifies the union of the body and blood, i.e. the quickening of the body with a soul, for blood is equivalent to soul."[11] For Jules Cashford, the Eucharist symbolises "the re-membering [of] the fragmented life of *bios*, which has continually to be reunited in the human heart with its eternal source, *zoe*."[12] As Spirit, wine may be seen as Sun, and bread, our daily comforting food, as Moon. So, the Eucharist may be said to consecrate the sacred marriage of *Sol y Luna*.

But what to make of the symbolism of the Eucharist being performed *on the Moon*? Of course, it may be argued that Aldrin is not a *bona fide* priest – his deed, therefore, does not carry the same weight as if a fully-fledged servant of God had been in charge. But if the symbolism is less potent, a magical operation it still is. Should we then see it as an act of conquest, the claiming of new land in the name of Christ? The Western political and religious pillars going at it again?

I would like to call on Sun/Moon symbolism to explore these questions, and Hugo Rahner, already mentioned, is particularly helpful in this regard. Which is a bit ironic, for Mr Rahner's tone is abrasively Christian, that is, downright dismissive and scornful of anything smacking of suspected paganism: at the bottom of ancient cults (a.k.a. pagans) "there is a genuine religious longing," he concedes, "but that longing expresses itself with a sort of drunkard's incoherence in a wild multiplicity of tongues, a multiplicity that still somehow contrives to achieve an effect of monotony."[13] Ouch. Here is a man not ready to compromise, and clearly vouching for the One And Only God. But, keeping in mind this unashamedly clear agenda, Rahner has also written very interesting pages on the Christian mystery of Sun and Moon.

Ritualizing the Last Supper, before Christ was taken away and put to death, the Eucharist is associated with the mystery of death and resurrection. For Rahner, the early Christian saw in Christ resurrected an image of the rising Sun: "the Christian of that day saw in the hero of Easter morning the sunlike Apollo-Helios who had slain the dragon Python."[14] In that sense, the Communion ritual perfectly matches the symbolism of *Apollo* on the Moon. Rahner goes further into Sun symbolism by stressing that "the meaning of Easter is that the life that began at Christmas is turned into life eternal":[15] that is, the 'Sun of Righteousness', who first arose at Christmas, secured immortality at Easter. "After Christ had completed the redemption of mankind through the sunset of the cross and the sunrise of Easter morning, he was free there and then to possess for himself the sunlight of his flesh glorified into eternity."[16]

But, as Cashford emphasized, the symbolism of Christ resurrected may just as well be associated with the New Moon reborn after three days of darkness. Easter is a soli-lunar day

– being the first Sunday following the first Full Moon after the Spring Equinox. There is rich symbolism here. Sunday is the day of the Sun. The day too, as already pointed out, of the *Apollo 11* landing. The Spring Equinox marks the midpoint in the course of the Earth around the Sun, with equal day and night, while the Full Moon is the moment when the Moon opposes the Sun and receives maximum illumination from Him. Campbell sees in this complex Easter dating an attempt to coordinate the lunar and solar calendars, and a symbol of the unity of life: "the dating of Easter according to both lunar and solar calendars suggests that life, like the light that is reborn in the Moon and eternal in the Sun, finally is one."[17]

Campbell was, however, going further, associating the mystery of Easter with the new life offered by the Space Age:

Easter and Passover are prime symbols of what we are faced with in the Space Age... We are challenged both mystically and socially, because our ideas of the universe have been reordered by our experience in space... The reality of living in space means that we are born anew, not born again to an old-time religion but to a new order of things. There are no horizons – that is the meaning of the Space Age.[18]

A new myth is called for, a new way of being in the world, a new cosmology: "the mystical theme of the space age is this: the world, as we know it, is coming to an end. The world as the centre of the universe, the world divided from the heavens, the world bound by horizons in which love is reserved for members of the in-group: that is the world that is passing away."[19]

By performing Communion on the Moon, Buzz Aldrin has drawn upon the powerful symbol of death and resurrection. It is a symbol that is significantly resonant with Moon symbolism. It is also, following Campbell, particularly appropriate to the new Space Age that is dawning. An old world died on the Moon, its soul resting on the mirror of our dreams. A new world is called forth.

Rahner sees a double Sun/Moon conjunction at play in the coming of Christ at Christmas, and in the end of times at Easter:

> The Sun of Christmas forms a union with the Christmas Moon and from this conjunction, both bridal and motherly, from this supernatural *synodos*, comes the procreation of all divine life for all the days to come... The redemption of the human race is a process that develops organically, a process in which the Church to the end of time must ever anew be giving birth and must ever be dying, for it will only be at the end of days that the whole Church will be truly flooded with the light of the Easter Sun. Then the Easter Sun will be united with the Easter Moon in a perpetual spring.[20]

The Sun/Moon conjunction at Christmas gives birth to divine life. Mary is the Christmas Moon and, having mothered the Sun of Righteousness, she has realized the sacred marriage: "as late as the fourteenth century," says Rahner, "a hymn to Mary sung by the French flagellants speaks of the supernatural conjunction of Christ the Sun and Mary the Moon."[21] Similarly, in a forgotten 13th century Latin manuscript: "When Sun and Moon unite; Darkness doth fly away; And all things grow more bright."[22]

For Rahner, the eventual realization of divine life will be symbolised by the Easter Sun/Moon conjunction between Christ, the Sun, and the Church, his Moon. But much work remains to be done:

> Christ the Sun sheds the simple primal light of his Father upon the Church which is his Moon, and the Church in her turn sheds it upon her new-born babes... but here below that light is not visible, for despite the new birth in baptism, all things in this life are still shrouded by the night of this

world.[23]

And so it will be until the end of time, when "all will be a beginning and all will be as the full Moon at Easter."[24] This the time of the Second Coming, "the day longed by the Christian as he wanders in the dark", "the day when Christ, the eternal Sun, and the eternal Moon, which is the Church, will give light to the innumerable host of the stars."[25]

Coniunctio

Can the contours of another symbolic language slowly emerge from behind the ritual of Communion? "By pronouncing the consecrating words that bring about the transformation," Jung pointed out, "the priest redeems the bread and wine from their elemental imperfection as created things. This idea is quite unchristian – it is alchemical."[26] Jung drew a strong parallel between the ritual of the Eucharist and the *Opus* of the alchemists, and between Christ and the philosophical stone:

> [In alchemy] what comes out of the transformation is not Christ but an ineffable material being named 'the stone', which displays the most paradoxical qualities apart from possessing *corpus, anima, spiritus* and supernatural powers. One might be tempted to explain the symbolism of alchemical transformation as a parody of the Mass were it not pagan in origin and much older than the latter.[27]

The alchemists sought to redeem matter by releasing the deity trapped in it – and this was the object of the *Opus* – an endeavour that Jung likened to "the same work of redemption which God himself accomplished upon mankind through the example of Christ."[28] So we may ask: behind the symbolism of Aldrin's Communion on the Moon, can we detect an alchemical process of transformation at work?

Indeed, by switching our gaze, we may see in the Moon landing by *Apollo*, rather than the solar annihilation of the Moon, the realization of the sacred marriage of Sun and Moon. Apollo the Sun, the Red King, meets His Divine Consort the Moon, the White Queen. After being plunged in darkness, we may start to notice the delicate contours of the New Moon slowly emerging, the birth of a new cycle – the fruit of the Sun/Moon conjunction.

"The alchemist," said Jung,

saw the essence of his art in separation and analysis on the one hand and synthesis and consolidation on the other. For him there was first of all an initial state in which opposite tendencies or forces were in conflict; secondly there was the great question of a procedure which would be capable of bringing the hostile elements and qualities... back to unity again.[29]

Sun and Moon, in their polarity, form a dualism, "either confronting one another in enmity or attracting one another in love."[30] "The alchemist's endeavours to unite the opposites culminated in the 'chymical marriage' between Sun and Moon (*coniunctio*), the supreme act of union in which the work reached its consummation."[31] Through their alchemical union, the Sun and Moon give birth to the Divine Child, the *lapis*, the third element that is greater than the sum of its parts since it contains both parents. By achieving this outer transformation, the alchemist, by sympathy, hoped to realize inner transformation – the inner Sacred Marriage. Jung stressed that the birth of the Divine Child took place in spiritual water, often referred to as 'the sea'. "The marvels of this sea," said Jung, "are that it mitigates and unites the opposites."[32] Was the *Sea of Tranquillity*, the point of landing of *Apollo 11* on the Moon, playing the role of spiritual water for the *coniunctio* to take place?

The planet Saturn played a key role in alchemy; it was the

most ubiquitous of the alchemical gods after Mercury. Its role was similar to Mercurius', facilitating the alchemical opus and the realization of the sacred marriage. Paracelsus held Saturn to be an intermediary principle between Sun and Moon. In alchemy as in astrology, Saturn and Moon have a strong relationship. In alchemy, Saturn was sometimes said to 'father' the Moon. In astrology, Moon and Saturn rule the opposite signs of Cancer and Capricorn: they are two poles of a unified principle, often referred to as the parental axis (Moon/Mother, Saturn/Father). Astrologer Liz Greene sees in this pairing a reflection of the "fluid maternal matrix of water [Moon] and the structured formality of earth, which 'moves by slow steps' [Saturn]."[33] Moon and Saturn move along the same rhythms, too. The Moon lunation cycle is 29.53 days on average (New Moon to New Moon), while Saturn revolves around the Sun in 29.45 years. One astrological technique, called secondary progressions, equates a day for a year – in that technique, the progressed lunation cycle is equivalent to the Saturn cycle. The affinity between the planets was further highlighted by Alan Leo, one of the founders of modern astrology. Leo echoed Paracelsus in assigning to Saturn an intermediary function between Sun and Moon: helping to align the lunar 'personality' with the solar goal of 'individuality'.[34] Or, in Leo's words: "So is the Moon, the representative of the personality, centred in Saturn, the planet controlling the path of discipleship or freedom from irresponsibility."[35] Saturn mediates between Sun and Moon: this is an interesting motif for indeed the giant rocket that lifted *Apollo 11* off the Earth was named *Saturn (V)*. So it looks like the wise old man, the *senex* archetype – Saturn – did participate in the *Apollo 11* opus in its customary mediatory role, linking Apollo the Sun with the surface of the Moon.

The alchemical union is the union of Spirit and Matter: "though the goal of the Opus," said Jung, "was undoubtedly the production of the lapis, there can be no doubt about its tendency to spiritualise the body."[36] This has also been referred to as the marriage of Heaven and Earth, a marriage that, moreover, "exhibits a numinous – 'divine' or 'sacred' – quality."[37]

Almost twenty years before the Moon landing – in the year 1950 – two major events took place, totally unrelated, that in retrospect I see as anticipating the 1969 Sacred Marriage on the Moon. On 3 June, the French successfully reached the summit of Annapurna[38] in Nepal, the first eight-thousander to be climbed. An important mark had been breached, sending humans higher up than ever before. The altitude above 8,000m is usually referred to as the 'death zone' due to the very poor oxygen content of the atmosphere. Then, man is hanging between life and death, reaching to the heavens at great risk. Five months later, on 1 November, through the papal bull *Munificentissimus Deus*, the Church dogmatically recognised that the Virgin Mary, "having completed the course of her earthly life, was assumed body and soul into heavenly glory." This is known as the Assumption of Mary, an event on which Jung placed great importance. Jung saw in the many Marian apparitions of the 19th and 20th centuries – most of them by children – the expression of an unconscious collective need for a female mediator to the Heavens: "One could have known for a long time that there was a deep longing in the masses for an intercessor and mediatrix who would at last take her place alongside the Holy Trinity and be received as the 'Queen of Heaven and Bride at the heavenly court'."[39] At a deeper level, this longing betrayed nostalgia for the sacred marriage, a rebalancing of the masculine and feminine poles: "it is psychologically significant for our day that in the year 1950 the heavenly bride was united with the bridegroom."[40] Campbell felt that the Assumption of Mary symbolically represented the Heavenly Earth: "Symbolically, the same tradition suggests it signifies the return of Mother Earth to the heavens, *the very thing that has occurred because of our journeys into space*" (my italics).[41] Another reading, however – one made by Baring and Cashford – held that this spiritualization of the body failed "because it [ended] up by depriving the body of substance."[42]

Quite clearly, I may be accused of stretching the symbolism to breaking point by 1. stitching the two 1950 events together, and 2. projecting them all the way to the 1969 *Apollo* mission, and I accept that. This is more an exercise in style than a literal demonstration. I especially acknowledge a rather loose symbolic take on the Annapurna climb but I could not resist it. It is hard not to see in the beautiful line of a mountain reaching for the sky a striving by Earth to meet the Heavens.

The Moon is transformation. She is the intermediate place where the dots may be joined. "In earlier times," says Cashford, "the Moon was Earth's star, a better Earth in Heaven."[43] Symbolically she was joining Heaven and Earth but then lost her status. She retained an intermediary role in the medieval cosmos but the cold gaze of science stripped her of this last role. Under the surface, though, she has kept the power to transform, even if we are not conscious of it. As Jung famously stressed, we moderns are terribly deluded in thinking that we have got rid of the gods. The old gods have not disappeared, they are just operating under a new guise. The fact that we do not recognise them any longer makes them all the more powerful. "The Moon stories," says Cashford, "suggest that the unconscious psyche yearns for transformation."[44]

The realization of the sacred marriage of Sun and Moon brings an unconscious content to consciousness: "becoming conscious of an unconscious content amounts to its integration in the conscious psyche and is therefore a *coniunctio Solis et Lunae*."[45] So we may ask: what have we become conscious of? Which new consciousness has been born of the *hieros gamos* on the Moon?

I want to argue that what has been revealed to us is a new vision of Earth. By the operation of the Moon landing, Earth, the blue marble, the lapis lazuli, the Divine Child of the *coniunctio solis et lunae*, has seemingly been rebirthed, bathed in a halo of intense numinosity. Seen as a Sun/Moon conjunction, the *Apollo* landing may then provide a mythopoeic foundation for a new planetary sensibility manifesting through the Gaia archetype. In these times of great need, when humanity stands at a crossroads, the mandala image of Earth has appeared on the screen of our consciousness, and this is anything but coincidental: "It may be that Nature," said esotericist John Michell,

> that great, enigmatic organism, directs in a subtle way the intelligence of her human products, reordering patterns of mind in accordance with the necessity of the situation. The present interests of Nature, it is now very clear, demand a fundamental change in the prevailing cosmology – our beliefs about the nature of the world and how best to relate to it.[46]

John Michell was the main figure behind the 'Earth Mysteries', a movement that investigated sacred features in the landscape, recognising an underlying orderly arrangement running along energy lines and specific points in the Earth – an arrangement that had been recognised by ancient cultures. This 'astroarchaeology', as it is usually called, runs counter to the prevailing dogma and is usually dismissed by mainstream archaeology. Michell's work is fascinating, though, and provides much insight into the ancient art of sacred archaeology, highlighting the constant preoccupation of the Ancients to marry Heaven and Earth through earthly constructions that would mirror the Heavens. Numbers and numerology played an essential part in this work:

> Ancient science was based, like that of today, on number, but whereas number is now used in the quantitative sense, for secular purposes, the ancients regarded numbers as symbols of the universe, finding parallels between the inherent structure of number and all types of form and motion.[47]

For the left hemisphere, obsessed with *What*, numbers cannot be anything else than quantitative. The right hemisphere, though, which is also concerned with *How*, acknowledges their qualitative and symbolic value.

The spontaneous emergence of a symbol to guide humanity in times of crisis is a recurrent motif in Western thought. Heinrich Zimmer thought that such symbols were manifesting through a Feminine image,[48] such as the Grail. Similarly, Jung held that the numerous sightings of flying saucers, with their round, mandala shapes, were expressing a longing for wholeness. For Michell, the image of the celestial city is such beacon. At the beginning of the Age of Aries, Stonehenge was built, he said. At the beginning of the Age of Pisces, St John had the revelation of the New Jerusalem. Today, at the beginning of the Age of Aquarius, I reckon that Earth is carrying the archetype of the Heavenly City:

> The traditional interpretations of such visions is that they presage the coming of a new dispensation that will reproduce on earth the harmonious order of the heavens. In the symbolism of all religions, a geometric construction representing the heavenly city or map of paradise has a central place. It occurs in sacred art as a mandala, a concentric arrangement of circles, squares, and polygons depicting in essence the entire universe. Related images include the labyrinth, the paradisial garden, the walled enclosure or temple precinct, the world tree, the enchanted castle on a rock, the sacred mountain, stone, well, or spring.[49]

John Michell paid a lot of attention to the squared circle, a classic template of sacred geometry. The squared circle is a circle drawn around a square in such a way that the circumference of the circle is equivalent to the perimeter of the square. While the circle is traditionally assigned to the Heavens, the square represents Earth. The squaring of the circle, therefore, symbolically represents the reconciliation of the opposites, the marriage of

Heaven and Earth. It is not only found in Stonehenge and in the blueprint of the New Jerusalem but also in cosmic geometry: "it is a remarkable fact of nature that the Earth and the Moon, placed tangent to each other and measured in miles, demonstrate the squared circle in terms of the very same numbers as make up the New Jerusalem diagram."[50]

It is quite challenging to convey Michell's thoughts without the help of diagrams. Suffice it to say that Michell is pointing to a remarkable symmetry between the sacred geometry of ancient temples and with the Earth/Moon duo:

> the earth's diameter is 7,920 miles, and the diameter of the moon is 2,160 miles. The combined radii of the two bodies are equal to 3,960+1,080=5,040 miles, and a circle with radius 5,040 has a circumference of 31,680 miles. That length, 31,680, is equal to the perimeter of the square containing the circle of the earth. It is also the measure of the squared circle in the New Jerusalem.[51]

These numbers carry a powerful symbolic charge. The number 1,080 – the length in miles of the Moon radius – is said to be a *yin* number:

> The number 1,080, the yin term in the cabbalists' equation of 666+1,080=1,746, is identified by all its symbolism with the Moon, the sublunary world, the waters below that are drawn by the Moon, the lunar influence on the earth's vital currents, the periods of the female, the unconscious, intuitive part of the mind, and the spirit that moves oracles.[52]

Michell worked with the ancient art of gematria, which consisted in assigning numbers to letters of the alphabet, each letter representing a particular type of universal energy. In Greek gematria, 1,080=the Holy Spirit=the Earth Spirit=Universal

Harmony=Prophecy=Wisdom but also equals 'Tartarus' equals 'Abyss': "it is the number of magic, imagination, and madness, and, above all, of that Mystery or principle of equivocation that lies at the heart of things and is not to be comprehended by any system of morality or rationalism."[53] In India the number of names in a litany of the Goddess is 108:[54] "when one finds numbers like 108... reappearing under several multiples in the *Vedas*, in the temples of Angkor, in Babylon, in Heraclitus' dark utterances, and also in the Norse Valhalla, it is no accident."[55] It also appeared in space: Yuri Gagarin spent 108 minutes in his maiden orbital flight around the Earth.

Contrasting 1,080 is the number 666, "the energy of the Sun and the principle of reason, will and authority... [It is the] generative power of the male, the call to action, the electric impulse that regulates the molecular field and gives form and order to chaos."[56] 666 has acquired a bad reputation by being associated with the Beast in Revelation but Michell argues that the number, as all symbolic numbers, displays positive and negative attributes: "the elemental symbol of 666 is the fiery flying dragon, the opposite of the earthbound serpent... Though its naked energy may be violently destructive, it cannot be described as evil, for it is the beloved mate of nature and the cause of life and beauty on earth."[57]

By dividing 1,080 by 666 we obtain 1.62 – the Golden Ratio. The addition of 1,080 and 666 amounts to 1,746, the so-called Number of Fusion: "it is the number by gematria of the Universal Spirit, which is the combination of the [negative and positive forces in nature]... [It] is the number of the fertilized seed from which, according to ancient philosophy, the whole universe grew up like a tree."[58] It is the number of the *coniunctio*, too, of the sacred marriage between Moon and Sun, and it has a strong association with Christ. 1,746=Son of the Virgin Mary=Grain of Mustard Seed=the Chalice of Jesus=Christ the Kingdom. Numerologist Tony Plummer noted how "the use of the Golden Ratio by early Christian writers [implied] a deep understanding of the cosmological forces operating within the universe."[59] "Specifically," he added, "the use of the number 666 to denote *mental* energy with active attributes, and the use of the number 1080 to denote *spiritual* energy with passive (or receptive) attributes, suggests an understanding of the universe as an interplay of opposing, but nevertheless complementary, forces."[60]

There is great poetry, and potency, in numbers. Pythagoras knew it, and Hypatia of Alexandria knew it. Our culture forgot about it. The sheer beauty of the cosmos is brilliantly revealed in its geometrical relations. John Martineau has written an enchanting book on cosmic harmony as revealed by numbers.[61] I highly recommend it. This cosmic harmony is particularly vivid in the various proportions linking Sun, Moon and Earth. Robin Heath, who has contributed much to the field of astroarchaeology, has noted many beautiful and uncanny 'coincidences' in the skies. For instance:

> 1/Sun=Moon and 1/Moon=Sun! In more detail 1/365.242 (the solar year)=0.0027379, which in days is 3 minutes and 56 seconds, the difference between sidereal and solar days, whilst 1/27.322 (the sidereal month)=0.0366, which in days is 52 minutes, the difference between lunar and solar days.[62]

Earth and Moon are in a relation of 3 to 11 – that is, the Moon diameter is 3/11th of the Earth diameter. They square the circle as 3:11. These numbers are reminiscent of the 3 astronauts on board *Apollo* 11.[63] As John Martineau remarks: "3:11 happens to be 27.3% and the Moon orbits the Earth every 27.3 days, the same period as the average rotation period of a sunspot. The Sun and Moon do seem very much the unified couple."[64] Martineau further notes that another celestial couple, Venus and Mars, displays a similar 3:11 vibration: "the Earth-Moon proportion is also precisely invoked by our two neighbours, Venus and Mars. The closest : farthest distance ratio that each experiences of the other is, incredibly, 3:11. The Earth and the Moon sit in-between them, perfectly echoing this beautiful local spatial ratio."[65]

Michell highlighted the key role of number 11 in the relationship between the three celestial bodies Earth, Moon and Sun: "the dimensions and distances of the earth, moon and sun relate to each other by simple ratios in which the number 11 is combined with the duodecimal numbers of the canon."[66] I once went on the Internet for a random search. I wanted to know a bit more about the esoteric significance of the number 11 and whether it could shed further light on the mission. I ended up on a website that read as such: "The number eleven carries a vibrational frequency of balance. It represents male and female equality. It contains both *sun energy* and *moon energy* simultaneously yet holding them both in perspective separateness. Perfect balance."[67] I had haphazardly stumbled upon a symbolic vindication of the *coniunctio* of the mission. Not that I knew what credence to give to such a source, but still, I was rather dumbfounded. Further digging seemed, however, to confirm that the number 11 uncannily realized the marriage of Sun and Moon. '11' is the bridge between the solar year and the lunar year: there are 12.368 lunations in one solar year, exactly 11 days more than in a lunar year of 12 lunations. Furthermore, 11 is the sum of 5 and 6, numbers which respectively carry lunar and solar attributes: "Psychic and lunar are the qualities of the pentagon, whereas the hexagon is rational and solar... An important exercise in sacred geometry is therefore to combine the hexagon and the pentagon in one synthetic figure."[68] This harmonisation is achieved through the Vesica Piscis, the almond-shaped figure at the intersection of two circles. Michell noted that the blueprint of the New Jerusalem involved a double Vesica, whose length was ten times the Number of Fusion, 1,746 – Sun and Moon united on Earth.

There is really no end to the magnificent interplay of the numbers at the heart of nature – numbers that have been replicated in sacred geometry. The geometry of Stonehenge, of Glastonbury Abbey and of the New Jerusalem – all based on the dimensions of the Temple of Solomon – mirror the geometry of Earth and Moon. The picture of Earth is a gift from *Apollo* to us all, the shining manifestation of a new consciousness, heralding a new way to be in the world. From the Moon, Earth appears as a symbol of wholeness, a healing mandala: a template of the Celestial City.

Chapter 10

Moon and Earth

Musings on the Squaring of the Circle

I believe in you, my soul, the other I am must not abase itself to you
And you must not be abased to the other.
– Walt Whitman

The Blessed Ones, truly! Nightingales, all, slowly bleeding on the edge of the frozen forest! *The Earth and the Moon square the circle.* What else do we need? How much more do we need? We bow to you, Mother Earth, and to your sister the Moon too, while the heat of the Sun weathers our backs. Does that sound New Agey? So be it, brother, and may the sound of trumpets drown your cynicism. And mine. Indeed we are turning into leather, brother! *A piece of the sacred bread went to the Moon.* Yes, and a piece of the sacred Moon came back to Earth. Our daily bread has gone cosmic. We have been fed ambrosia, and it gently curled up in the one-sixth gravity of the Moon. Divine nectar in slow motion. This is so liberating! In Chartres, death and rebirth under the oak. Under the oak, yes! The labyrinth is a giant worm coiled around the World Tree. Death and rebirth! Our gaze has shifted, and we *see.* The seeds of a New Moon have been planted. Souls were dancing around Armstrong while he was looking back at Earth. Ghosts were spreading flowers under Aldrin's feet. They drew a Vesica around the *Eagle* with dusty bones, and they sang! Dear One, how much they sang! They brew moon dust with sunrays in an apocalyptic cauldron and fumes escaped to the dark heavens of yesteryear. Apollo spoke with a terrible voice, a voice of thunder, of moonquakes. Bees escaped from His golden mouth! Apollo=1061=Jason=Jonas. 1061 is the 178th prime number. Does that matter? In the Bible, Job 18.6, 1061 = "the light shall be dark in his tabernacle and his candle shall be put out with him". Dark light, Midnight Sun. On her silver tripod, the Pythia shouted at the top of her lungs in the void of space. She was heard far, far away, by tall, thin beings of light dwelling bare-chested on floating cities. They, too, once witnessed the Sun setting on their decaying planet, and then they were reborn! Why? Because they found *faith*! 1772=The Spirit of the Sun=The Soul of the Moon. Wassail! Symbolism, truly, is worthless, and so is poetry – unless it brings the whole heart to sing! And beauty to boot! What is beauty? A butterfly on the brink of asphyxia, is my two pennies. We are so lacking in beauty. We, the scavengers!! Ripping apart smoky carcasses with our obsidian teeth. Extracting the sacred juice from old wounds. We despair at the morbidity around us, at the engineered scarcity, at the insane obliteration of life. And now bright brains want to teach compassion to computers, Goddammit! Our *hubris* knows no bounds! We lost our hearts long ago and we are well on the way of losing our minds as well. What are we going to be left with? Artificial DNA and synthetic skin, that's what! Virtual paradises and PC cyborgs with bogus smiles. Fake, fake, all fake! "The heart as lion is truly king of beasts, a bestial King, and our inner beauty – our dignity, nobility, proportion, our portion of lordliness – comes, as lore of character has always assumed, from the animal of the heart."[1] Thus spoke Hillman. Let us free the beast from the cage! Let us roar with lions! Let us dance with the gypsies under the moonlight! Django Reinhardt lost the use of two fingers in the fire of his caravan. Did that prevent him from being the greatest guitarist ever? Hell, No! He was a lion and he knew beauty! When I look at the *Blue Marble* picture taken by *Apollo 17*, I know that I am in the presence of the divine. Does that offend you? Why does that offend you? I come in peace, open palms, garland of flowers round the neck. What do I mean by 'divine' anyway? *'Ah!'* may be a better word. Or

'Awe'. Or *Huh, Hey, How, Who. Ho Ho Ho*? The last laugh is always for Father Christmas, reindeers floating high on fly agaric. The curtains are ripped apart by chillum-smoking-red-dwarves. And all the while, wet and soaked under the dripping stars, Tom Bombadil was busy playing the sitar under the banyan. The Holy Fool! Humility, compassion, wholeness crash the party bearing gifts – the true magi of a new dawn, riding their ecstatic camels! The djinns were wearing jeans, would you believe it. When the wind blows from the North, you may see me scry on water. On the surface of ancient lakes, I can slowly see you, sister, wiping your brow in the heat of the battle. You lose your arm, then another, and your legs give way under the weight of iron skies. But you believe in the Great Wheel and you nod to the White Faced Lady of the Twenty-Nine Wounds – and all of a sudden you grow back, stronger than ever before, wind howling through your hollow bones. Up there, atop the jade mountain, alone and naked and rapturous like Teresa, you are pierced by the thousand arrows of *Nwyfre*, the life force. Ivy grows in your mouth. I bet St Thomas does not know how many angels can land on your diamond skull! But no doubt he will spend hours debating this with his buddies! And here you are, hissing like the Grand Mother of the Sacred Tribe. By Jove, she is so hungry this Grandma, swallowing bubonic toads like there is no tomorrow! HAIL TO YOU, SISTER! You dance on crop circles, and your feet barely touch ground. But hey! The music of the spheres is now available on Spotify for $9.99 a month. Bargain! Hail to you, Radiant One! I can feel my heart expanding, as if I carried the Earth herself in my chest. I am the creative juices at boiling point, a mammoth kettle running on Elmo's fire. This chapter contains exactly 1111 words. I sense a giant pulse rippling in translucent waves throughout the universe. I am aboard *Apollo 14*, one cell beating with myriads of cells, and I see Edgar Mitchell glued to the porthole, his veins about to explode in a kaleidoscope of shapes and forms and colours. He is singing the mantra of

bliss for the Earth beckons. Om namah shivaya! "We are too late for the gods, and too early for Being," said Heidegger. There is much work to do, and we are in dire need of boddhi satvas.

Chapter 11/1

Gaia: *The Rebirth of the Goddess of the Earth*

The space programme, which was meant to show mankind that its home was only its cradle, ended up showing that its cradle was its only home.
– Robert Poole

Almost one hundred years ago to the day, German philosopher Rudolf Otto published a very influential essay titled *The Idea of the Holy*. Otto felt that the concepts of 'holy' and 'holiness' had lost some of their original potency: while they still carried a strong ethical sense, a connotation of moral goodness, they had lost "an overplus of meaning" that they once conveyed. Otto was at pains to qualify this overplus of meaning and eventually coined the word 'numinous', derived from the Latin word 'numen'.[1] The numinous was an emotional experience, one that could not be qualified but only shared: "it cannot, strictly speaking, be taught, it can only be evoked, awakened in the mind; as everything that comes 'of the spirit' must be awakened."[2] At the heart of numinosity was the *Mysterium Tremendum*, "that which is hidden and esoteric... beyond conception or understanding, extraordinary and unfamiliar."[3] Through the *Mysterium Tremendum*, said Otto, we encountered the notions of awe-fulness and fascination. We came face to face with the 'wholly other':

> The concepts of the 'transcendent' and 'supernatural' become forthwith designations for a unique 'wholly other' reality and quality, something of whose special character we can *feel*, without being able to give it clear conceptual definition.[4]

The numinous is beyond language, escaping the attempts of the rational mind to corner it. It is a felt experience, with a sense of the ineffable, and it carries a strong emotional charge. One might say that is replete with *Sereti* – a commanding presence that imposes respect. The concept of numinosity has proved very popular. Otto had obviously tapped into a deep collective need. The *Mysterium Tremendum*, exiled from Western shores for too long, was screaming for recognition. 'Numinosity' became a staple word for depth psychologists. Jung used it to describe the emotional component accompanying deep unconscious projections: "the projection of a highly fascinating unconscious content... exhibits a numinous – 'divine' or 'sacred' – quality."[5] By assigning a divine quality to a psychological content Jung was suspected of psychologizing the Gods. Where indeed does the psyche end and where does God begin? Saiva Tantrism would say that the individual psyche is Consciousness-in-action, God expressing Herself in particular, limited ways so as to experience Her infinitely varied nature, but that the psyche is an illusion, a distraction, an obstacle to experiencing our true divinity. In a 1959 BBC interview, Jung was asked if he believed in God. "I do not believe," was the remarkable answer, "I know." This devilish, tricksterish reply seemed to imply that far from kicking the Gods off their Olympus, Jung was in fact wining and dining at their table. Behind the archetype of the Self, however, Jung always refused to venture, considering that it was beyond his remit as a psychologist. What was behind the psyche was a matter of faith – or maybe, judging by his response, of 'knowing'.

The picture of Earth brought back by *Apollo 8*, as well as the subsequent pictures taken by later missions – especially *Apollo 17*'s *Blue Marble* – exhibited a numinous quality. Seen from space, Earth has come shrouded in a sacred veil. Something mysterious has been awakened – the *Mysterium Tremendum* brought to life – and the rational mind has been found wanting, unable to express our emotion. It is then that the old gods and

227

goddesses, still alive in the deep recesses of the human psyche, come to the rescue. The 'highly fascinating unconscious content' that has been projected on to Earth has indeed called forth an ancient divinity to express the *felt* numinosity of the Earth: an archetype that conveyed in symbolic language our sense of awe and wonder, a goddess who carried an ancient wisdom buried in the sands of history. The summoning of that divinity, however, came from a rather unexpected corner: no less than science.

In 1961, British scientist James Lovelock was hired by NASA to analyse the composition of extraterrestrial surface atmospheres. More specifically, anticipating David Bowie's 1971 song, his work focused on probing the possibility of life on Mars. Comparing Earth's and Mars' respective atmospheres, Lovelock was struck by the incredibly stable yet complex atmosphere found on Earth. By contrast, Mars was essentially made of carbon dioxide and "showed no sign of the exotic chemistry characteristic of Earth's atmosphere."[6] What surprised him most was the remarkable capacity of Earth for homeostasis – that is, her ability to maintain steady internal conditions. The content of oxygen in the atmosphere, for instance, had been stable for hundreds of millions of years at 21%, the exact right number for life to prosper. A few percentage points lower would leave life forms gasping for air, a few percentage points higher and oxygen, a highly combustible gas, would consume everything in a ball of fire. Lovelock intuitively sensed that the chemical composition of a planet's atmosphere was intimately linked to the presence of life: "our results convinced us that the only feasible explanation of the Earth's highly improbable atmosphere was that it was being manipulated on a day-to-day basis from the surface, and that the manipulator was life itself."[7] Lovelock further used the examples of methane and oxygen to illustrate his point. In sunlight, he pointed out, the two gases react to create carbon dioxide and water vapour. In order to keep the Earth's atmospheric mixture constant, 500 million tons of methane and twice as much oxygen have to be added to the atmosphere every year. This, said Lovelock, was "improbable on a biological basis by at least 100 orders of magnitude."[8] Lovelock proposed therefore that, in a virtuous feedback mechanism, "the atmosphere... [acted] as a dynamic extension of the biosphere itself."[9] In other words, rather than adapting to pre-established planetary conditions, life was actively involved in creating the conditions to sustain itself. Initially restricting 'life' to living organisms, Lovelock later theorized that mountains, volcanoes, rocks, forests and oceans also played an essential part in the regulation of the Earth's atmosphere. All this was more than mere intuition. Lovelock created a tool, the Daisyworld model, through which the theory could be mathematically tested with predictable results. Lovelock's view implied a holistic notion of the Earth: a giant organism that self-regulated so as to create the best conditions for its own survival. Earth was an "entity... [whose] properties... could not be predicted from the sum of its parts."[10] Lovelock's work belonged to the mathematical field of systems theories that, with the advent of computerisation, was gaining much traction in the seventies – in particular through chaos theory and the iconic Mandelbrot sets. The Earth was described as a "cybernetic system that [sought] an optimal physical and chemical environment for life on this planet."[11] Yet, despite adhering to strict scientific orthodoxy, Lovelock quickly became highly suspect for his more conservative colleagues.

Notwithstanding the safe objective language of science, Lovelock was indeed looking for a name for his iconoclastic theory – a name that would resonate with and speak to people's hearts. Enter author William Golding, of *Lord of the Flies* fame. Golding was living in the same village as Lovelock and, on one of those lazy afternoon strolls that make history, suggested to use the name Gaia, after the Greek Earth goddess, to christen the new theory. Lovelock was immediately won over:

In spite of my ignorance of the classics, the suitability of the choice was obvious. It was a real four-lettered word and would thus forestall the creation of barbarous acronyms, such as Biocybernetic Universal System Tendency/Homeostasis. I felt also that in the days of Ancient Greece the concept itself was probably a familiar aspect of life, even if not formally expressed.[12]

Thus was born the Gaia Theory, which was to deeply affect our understanding of Earth as well as seriously question our very role as Earthlings. For indeed, Lovelock's scientific theory went way beyond the confines of science. By recognising that man was one strand in the community of life, the Gaia hypothesis had profound moral implications: "I began to see us all, as part of the community of living things that unconsciously keep the Earth a comfortable home, and that we humans have no special rights only obligations to the community of Gaia."[13]

After millennia of being hidden away from view Gaia was reborn, and one cannot but not notice the synchronicity of her resurgence with the *Apollo* pictures of Earth. By summoning the old Goddess, Lovelock was, however, stepping into tricky territory. Myth and science do not make easy bedfellows in a world gripped by the left hemisphere: "The critics took their science earnestly and to them mere associations with myth and storytelling made it bad science."[14] Lovelock was faced with a conundrum, trying to appeal to both hemispheres. "To establish Gaia as a fact I must take the... path... of science. As a guide on the best way to live with the Earth, it will only be believed if it comes with majority support from the scientific community – politicians and governmental agencies that do not act on myth – and demand scientific approval." With hindsight, one has to severely question this assumption. Despite the work of IPCC,[15] man-made climate change is still unrecognised by many segments of the population and by governments stuck in an ideology of consumerism. "The American way of life is not up for negotiation. Period," declared President George Bush at the onset of the 1992 UN Conference in Rio, and no amount of scientific evidence was ever going to change that. Maybe aware of the risk inherent in appealing to mind only, Lovelock felt that another language was also required to touch hearts: "there is no betrayal of Gaia, we need the restraint of scientific conduct for investigation and theory testing and we need the poetry and emotion that moves us and keeps us in good heart while the battle goes on."[16] This tension between scientist and myth teller, which Jung also grappled with all his life, proved quite challenging to Lovelock, who became increasingly uncomfortable with the New Age appropriation of the Gaia figure and the implied view of Earth as a living, animate, conscious being:

The idea of Mother Earth or, as the Greeks called her, Gaia, has been widely held throughout history and has been the basis of a belief that coexists with the great religions. Evidence about the natural environment accumulates and the science of ecology grows. This has led to speculation that the biosphere may be more than the habitat of all living things. Ancient belief and modern knowledge have fused emotionally in the awe with which astronauts with their own eyes and we by television have seen the Earth revealed in all its shining beauty against the deep darkness of space. Yet, this feeling, however strong, does not prove that Mother Earth lives. Like a religious belief, it is scientifically untestable and therefore incapable in its own context of further rationalization.[17]

James Lovelock was born under a Dark Moon – a time when the old makes way for the new, when a cycle ends and another opens up. A time when visions are born. But there is more. In Lovelock's astrological chart the Sun is placed in his own sign, Leo, close to the Moon in her own sign, Cancer – two very strong placements. This book argues that the Earth can be symbolically seen as the Divine Child of the Sun/Moon conjunction. The natal horoscope of James Lovelock, the father of a 'new' vision of Earth, vividly illustrates this point. Sun and Moon, in their own signs, unite and give birth to Gaia, our Earth.

Are we moderns turning a corner? Are we about to re-establish a new covenant with Earth, one that is calling forth an ancient sensibility? For the 'new' vision of Lovelock is, as is so often the case, nothing other than old wisdom clothed in modern scientific garb. "The position of the Earth archetype as an expression of man's relation to the earth," said Neumann, "is naturally, in the final analysis, an expression of his relationship to his Earth and his Heaven."[18] Indigenous traditions have always carried a deep reverence for the Earth, constantly engaged in an I-Thou dialogue. It was modern man that broke this dialogue and Neumann squarely put the blame for this rupture at the feet of a Masculine principle run amok – a bloated solar consciousness:

> Devaluation of the Earth, hostility towards the Earth, fear of the Earth; these are all from the psychological point of view the expression of a weak patriarchal consciousness that knows no other way to help itself than to withdraw violently from the fascinating and overwhelming domain of the Earthly. For we know that the archetypal projection of the Masculine experiences, not without justice, the Earth as the unconscious-making, instinct-entangling, and therefore dangerous Feminine.[19]

Indian sage Sri Aurobindo had a similar take: "The will to escape from a cosmic necessity because it is arduous, difficult to justify by immediate tangible results, slow in regulating its operations, must turn out eventually to have been no acceptance of the truth of Nature but a revolt against the secret, mightier will of the great Mother."[20]

And yet one should refrain from idealizing indigenous cultures, for there too lurks a shadow. In *The God Delusion* Dawkins discusses the work of Kim Sterelny, a philosopher of science who worked with aboriginal tribes in Papua New Guinea:

> The very same people who are so savvy about the natural world and how to survive in it simultaneously clutter their minds with beliefs that are palpably false and for which the word 'useless' is a generous understatement... They combine [says Sterelny] "a legendary accurate understanding of their biological environment with deep and destructive obsessions about female menstrual pollution and about witchcraft. Many of the local cultures are tormented by fears of witchcraft and magic, and by the violence that accompanies those fears". Sterelny challenges us to explain "how we can be simultaneously so smart and so dumb".[21]

While I find it hard to condone the terms used by Dawkins and Sterelny–'false', 'useless', 'smart', 'dumb' – I think that they raise a valid point. One understands the Enlightenment philosophers, so dead intent on raising barriers against superstition and religious manipulation. But have we thrown the baby with the bathwater? "The attempt to deny or stifle a truth," said Aurobindo, "because it is yet obscure in its outward workings and too often represented by obscurantist superstition or crude faith, is itself a kind of superstition."[22]

One might say that the aboriginals of PNG were living in a world ruled by the Great Mother. French anthropologist Lucien Levy-Bruhl coined the term *participation mystique* to describe the mentality, based on a *felt* participation, of the indigenous tribes he was encountering. Although strongly coloured by the cultural bias typical of the 19th century,[23] there is much to like in this groundbreaking work. Levy-Bruhl was fascinated by indigenous mental processes, and in particular by the ability to accept a double reality, which he variously described as "identity, consubstantiality, sympathy, solidarity, duality-unity, etc."[24] A classic example of such duality-unity is given by the photograph. The resistance of indigenous people to have their picture taken is legendary. From a Western detached perspective, it sounds like poor superstition. But Levy-Bruhl took great lengths to point out that for such mentality, "the photograph is felt as *being* the very person of whom it is the portrait."[25] It is both different, and one and the same. Similarly, "the footprint in the sand [is felt] as being the very animal which has escaped; the leopard in the jungle as being the very Naga who lives in the neighbouring village. The ideas of 'one' and 'two' play no part here in 'primitive man's thought'."[26]

Levy-Bruhl struggled all his life to translate into language this felt participation that is never expressed by indigenous people and is in itself alien to conceptualisation, although crucially it is not alien to logic. Indeed, Levy-Bruhl was at pains to correct his earlier theory of a pre-logical mentality among indigenous people – that is, a mentality operating before, or outside, logic – emphasizing later on that indigenous people had the same logical mentality as moderns but that this mentality was cohabiting, indeed was often subsumed, by what he called a mystical mentality which rendered secondary causes irrelevant. Therefore *a physical* impossibility did not necessarily mean a *logical* impossibility – a point that moderns have immense difficulties to accept. As Levy-Bruhl said, "what disturbs us is the violence done to … our mental habits."[27] This mentality, in which a person can be both dreaming in her bed and *simultaneously* physically visiting the next village, is unbearable to the left hemisphere, which needs 'either/or' answers. Duality-unity is not bad superstition, nor the sign of a weak, sick or backward mind. It is another way to engage with reality, and it has much to teach us. Although obliterated by the philosophy of the Enlightenment, remnants of it are still being gathered like a treasure-trove by poets, shamans and diviners. Indeed the whole field of divination, including astrology, operates along duality-unity, and this is where it stands irremediably at odds with the modern mind. It is also, it must be noted, at the heart of the sacrament of the Eucharist, when bread and wine become *consubstantial* with the body and blood of Christ.

But one has also to recognise the potential pitfalls of *participation mystique* – the uncritical acceptance of ancestral lore, sometimes engendering terrible suffering, the potential scapegoating, the unconscious projections, the burden of an overbearing Great Mother, the coercive grip of the collective, the alienating nature of superstition and obscurantism. As Sri Aurobindo pointed out, "it became necessary for a time to make a clean sweep at once of the truth and its disguise in order that the road might be clear for a new departure and a surer advance. The rationalistic tendency of Materialism has done mankind this great service."[28] Needless to say, a lot of these unfortunate traits are also found in our modern, 'enlightened' societies, but the point here is not to wag the finger. Rather, we must try and understand what *participation mystique* is, and what it can offer or not to a world whose utter rejection of it has increasingly been casting its own shadow. Contemporary men and women must re-enter into an I-Thou dialogue with Gaia, yet even this is not enough. With *participation mystique* the boundaries between *I* and *Thou* were blurred. Further on, as consciousness evolved, *I* came to recognise *Thou* as separate yet as worthy of respect and reverence. It was modern man who denied *Thou* any agency and transformed it (sic) into an object, an *It*. We must now not only reverse this movement but also engineer a conversation in which we are able to hold *I* and *Thou* as simultaneously identical *and* distinct. We must recognise that we are an integral part of the Gaian community but also, with the blessing and the curse of a self-reflexive consciousness, separate from it. Duality-unity: this, to me, is the gift of *participation mystique*. A new way to re-engage with the world, in which we are both in and out. It is a paradox that only the Imagination can handle.

Barbara McClintock had Imagination in abundance. Not the imagination that makes you dream of lying on a beach in the Bahamas, No. Imagination as in 'duality-unity'. McClintock was a brilliant and eccentric biologist and she made her life work to study corn. Neglected for decades because of her unconventional methods, her work was eventually recognised, earning her the Nobel Prize in 1983 at the respectable age of 81. In her work she managed beautifully to combine intuition with reason. She had what she called a "feeling for the organism". As her biographer Evelyn Fox Keller put it, she was adamant that "one learns by *identification*"[29] (my italics). Spending a lot of time in the fields, McClintock developed a deeply *felt* understanding of the corn plant: "I start with the seedlings, and I don't want to leave it. I don't feel I really know the story if I don't watch the plant all the way along. So, I know every plant in the field. I know them intimately, and I find great pleasure to know them."[30] Here is the language of *participation mystique* again, but one of a different kind. For after gaining her intuitive insights, McClintock was donning the mantle of the scientist to draw the conclusions of what she had intuited. She shared with quantum physicists Niels Bohr and Erwin Schrödinger a deep reverence for Buddhist and Hindu spiritualities, which led her to an absolute conviction that all life was one: "Basically, everything is one. There is no way in which you draw a line between things."[31] She would go as far as calling herself a 'mystic' – a fraught word in scientific circles, one that Lovelock, for instance, would not touch with a bargepole – and she was not shy in denouncing the limits of a scientific method predicated on separating and compartmentalizing. Not the best way to earn the favours of your (male) colleagues: she duly paid for transgressing the unwritten code of science etiquette, being marginalised for decades. Her main paradoxical insight was that science came *after* the knowing, not the other way round. But she was utterly baffled by the source of this knowing:

Why do you know? Why were you so sure of something when you couldn't tell anyone else? You weren't sure in a boastful way; you were sure in what I call a completely internal way… What you had to do [then] was to put it into their frame… so you work with so-called scientific methods to put it into their frame *after* you know. Well, [the question is] *how* you know it.[32]

Was this what Jung meant by his "I know"? Barbara McClintock was a truly remarkable woman, perfectly embodying what is being asked of us collectively: the ability to think with both left and right hemispheres, the joining of reason with intuition, the deference to the Imagination, and the crucial acknowledgement of, and primacy given to, the right side of the brain. Then the Master recovers her throne and the Emissary his role. "Her 'feeling for the organism' is the mainspring of her creativity. It both promotes and is promoted by her access to the profound connectivity of all biological forms – of the cell, of the organism, of the ecosystem."[33] Inevitably this deep feeling of oneness led McClintock to comment on our disastrous disconnection from Earth:

We've been spoiling the environment just dreadfully and thinking we were fine, because we were using the techniques of science. Then it turns into technology, and it's slapping us back because we didn't think it through. We were making assumptions we had no right to make. From the point of view of how the whole thing actually worked, we knew how part of it worked… We didn't even inquire, didn't even see how the rest was going on. All these other things were happening and we didn't even see it.[34]

The fact that it was a woman who so beautifully showed us the way forward is, in my view, anything but a coincidence. As the French poet Louis Aragon put it: *"La femme est l'avenir de l'homme."*[35]

No longer our Mother, the Earth has lost her numinosity, only to recover it through the *Apollo* images. It is the merit of Lovelock to have summoned back the image of Gaia, even if he himself eventually felt ambiguous about it. The genie was out of the bottle, however, and there was no turning back. "It is no coincidence that the twentieth century returns to the Greek mind," say Baring and Cashford, "for in the West the last Goddess of Earth was Gaia."[36]

Who then was Gaia and what was her role in Greek mythology? The Homeric hymns – always a good place to start – extolled her abundance and life-giving properties, but also recognised that it was her privilege to take away what she had given:

HOMERIC HYMN TO GAIA, MOTHER OF ALL[37]
Gaia, mother of all,
I shall sing,
The strong foundation, the oldest one.
She feeds everything in the world.
Whoever walks upon her sacred ground
Or moves through the sea
Or flies in the air, it is she
Who nourishes them from her treasure-trove.
Queen of Earth, through you
Beautiful children
Beautiful harvests,
Come.
It is you who give life to mortals
And who take life away.
Blessed is the one you honour with a willing heart.
He who has this has everything.
(...)

He who has this has everything. These ancient wise words seem to have been buried deep into the sands of history. For the Greeks, Gaia was the original Deity, the first one to emerge from Chaos. Out of herself she gave birth to Ouranos, the Sky, and in the classic tradition of son-lovers of the Goddess, she mated with him and gave birth to the next generation of gods and goddesses, the Titans – amongst which we find Selene, the Moon, and Helios, the Sun. An extraordinary loop is closed: Gaia originally gave birth to Sun and Moon and in turn, when She was all but forgotten, they reciprocated and birthed a new vision of Her.[38] The cycle is complete and yet Gaia today is not Hesiod's Gaia. As TS Eliot so evocatively put it in *Little Gidding,*

We shall not cease from exploration
And the end of all our exploring
Will be to arrive where we started
And know the place for the first time.
Through the unknown, remembered gate
When the last of earth left to discover
Is that which was the beginning.

Are the roads leading back to Delphi, "the last of earth left to discover, which was the beginning"? For Gaia was the first ruler of Delphi and, as Baring and Cashford point out,

the later Gods of Delphi – Poseidon, Dionysus and Apollo – never forgot that she was 'there' first. The priestess of Apollo, the Pythia… would always open the Delphic rituals with an invocation to Gaia:
I give first place of honour in my prayer to her
Who of the gods first prophesied, the Earth.[39]

In myth, Apollo and Gaia are powerfully related through their rulership of Delphi. They are also family: as the son of Zeus, son

239

of Kronos, son of Ouranos, Apollo is Gaia's great-grandchild. Millennia after taking over Delphi, Apollo, in a reverse move, defers back to Gaia, the original matrix of Creation, his great-grandmother – a neat counterpoint to the original murder of Tiamat by her great-great-great-grandson, Marduk, the first solar myth.[40] Maybe we have learned something along the way if young males, rather than killing or despoiling their female ancestors, honour them – the ancient feminine principle revered again! Revealing Earth in her vibrating numinosity, *Apollo* indeed has offered a new vision of Gaia to carry us forward in these times of great need.

Campbell thought that the astronauts had "pulled the Moon to Earth and sent the Earth soaring to Heaven."[41] This is a neat image but isn't an Earth in Heaven at risk of dematerializing Earth, of stripping her of her body – in fact of reinforcing the spirit/matter split? Baring and Cashford thought so with regards to another symbol of a heavenly Earth, the Assumption of Mary. In dogma, the Mother of God is not a goddess, they asserted, and therefore "if Mary is *experienced* as divine – which in the old days was called a 'goddess' – then her ascent from earth to heaven must subtly diminish earth, that is to say, the sacredness of earthly life, which is the possibility of the divine immanent in creation."[42] In their view, the Assumption of Mary spiritualized the feminine. It acknowledged matter from the point of view of spirit and was therefore incomplete and partial. And so, they added, "the reinstatement of the feminine principle fails because only part of the feminine principle is acknowledged, and so the split is perpetuated, or perhaps even worse, enshrined."[43] A similar objection may be raised for Campbell's image of an 'Earth soaring to Heaven'.

The split between spirit and matter, said Aurobindo, could only end in either the death of God or the estrangement from Nature: "If we assert only pure Spirit and a mechanical unintelligent substance or energy, calling one God or Soul and the other Nature, the inevitable end will be that we shall either deny God or else turn from Nature."[44] Both propositions have turned out to be correct. In Freudian terms, we have both killed God the Father and raped Earth the Mother. Yet for Aurobindo, Earth was the very ground where Spirit met Matter:

> The affirmation of a divine life upon earth and an immortal sense in mortal existence can have no base unless we recognise not only eternal Spirit as the inhabitant of this bodily mansion, the wearer of this mutable robe, but accept Matter of which it is made, as a fit and noble material out of which He weaves constantly His garbs, builds recurrently the unending series of His mansions.[45]

The unending series of His mansions. How beautiful. This reminds me of Garcia Lorca: "One must awaken the duende in the remotest mansions of the blood."[46] Duende: earthiness, death, a dash of the diabolical. Garcia Lorca was a great admirer of the Spanish poems of deep song, deeply rooted in Earth, "magnificently pantheistic: the poets ask advice from the wind, the earth, the sun, the moon, and things as simple as a violet, a rosemary, a bird":[47]

> Only to the Earth
> Do I tell my troubles
> For nowhere in the world
> Do I find anyone to trust.

The unending series of His mansions. Each and every creature, each and every rock, tree, drop of water, a nest for Her infinite grace. The Kingdom of God is all over the Earth and men do not see it (Gospel of St Thomas). "Cleave the wood, and I am there," said Jesus. "The Promised Land," said Campbell, "is a corner in the heart, or it is any environment that has been mythologically spiritualized."[48]

Like many indigenous traditions all over the world, the Navaho were living in the Promised Land: "Living in a desert, the Navaho have given every detail of that desert a mythological function and value so that wherever persons are in that environment, they are in meditation on the transcendent energy and glory that is the support of the world."[49] This is the gaze that we have to recover if we are to renew our bond with Earth. Arguably, the *Apollo* pictures have greatly contributed to open our eyes to "the transcendent energy and glory" that is Gaia herself. But this holistic vision has also to be translated into particulars. In Greece, said Cashford,

> Gaia was *both* the name of the Great Mother Goddess Earth *and* the everyday Greek word for earth... The 'Earth that gives us grain' was also the 'Mother that feeds the world', who was also 'the oldest one, the Mother of the gods'. So 'Gaia' as Goddess, ground and globe, was always transparent to the deeper poetic vision.[50]

The earth under our feet, the divine Earth, 'the bride of starry Heaven', was ensouled:

HOMERIC HYMN TO GAIA, MOTHER OF ALL[51]
Continued
(...)
He who has this has everything.
His fields thicken with life-giving corn,
his cattle grow heavy in the pastures,
his house brims over with good things.
The men are masters of their city,
the laws are just,
the women are fair,
great riches and fortune follow them.

Their sons delight in the ecstasy of youth,
Their daughters play
in dances garlanded with flowers,
they skip happily on the grass
over soft flowers.
It was you who honoured them,
sacred goddess, generous spirit.
Farewell, mother of the gods,
bride of starry Heaven.
For my song, allow me a life
my heart loves.
And now I shall remember you
and another song too.

Chapter 11/2

Gaia: *Putting earth into Earth*

Hope for a renewal of the creative forces of the planet lies in a realization that the Earth is primary and that humans are derivatives.
– Thomas Berry

We moderns are now invited to make again the connection between globe and ground – nay, more than that, we are *requested* to make it, for our own good and for the good of the community of life that is Gaia. Praise be to Lovelock for his stroke of genius. By ensouling the world, the poetic image of the Goddess does indeed help us bring about a renewed vision of Earth.

This ensoulment of Earth finds its polar opposite in the image of the Waste Land, which stands as a symbol of a wounded Masculine and a dishonoured Feminine:

What are the roots that clutch, what branches grow
Out of this stony rubbish? Son of man,
You cannot say, or guess, for you know only
A heap of broken images, where the sun beats,
And the dead tree gives no shelter, the cricket no relief,
And the dry stone no sound of water.[1]

The Grail legends encountered enormous success in Europe in the 12th and 13th centuries, at a time of profound questioning of the old feudal model. "As the patristic order weakened," said philosopher Paul Zweig, "and the social imbalance of feudalism caused a disruptive mood to prevail on all levels of society, desires that had been half repressed and half-forgotten began to find conscious expression."[2] The Waste Land symbolised this barrenness, this imbalance. The King – the Masculine – was wounded in the groin and could no longer perform his fertilizing role. The earth was dry. Only by finding the Grail – an image of the Feminine – could the land be healed. The image of the Great Mother reappeared in these times of great need:

> In the twelfth century, a new mood is voiced simultaneously in many parts of Europe: a need to reaffirm qualities of experience that had long been neglected. The resurgence of mysticism, the rediscovery of the old Gnostic heresies, the polite rebellion of *fin'amor* in Provence are cited as examples of a new breakthrough of the Great Mother into the mind of authoritarian Europe.[3]

Being essentially a masculine quest, the Grail Quest still placed men at the centre of culture. Yet – a rare feat in the last four thousand years – the Feminine principle was held in high esteem and extraordinary female figures emerged, like Eleanor of Aquitaine and Marie de Champagne, who beautifully embodied this principle.

The Sixties, and by extension the early Seventies, witnessed a resurrection of this spirit. The Great Mother reappeared through a Gaian consciousness that found manifestation in the Gaia Theory, in the *Apollo* pictures, in Deep Ecology, in the Goddess movement, in the creation of Friends of the Earth, of Greenpeace and of Earth Day, in neo-shamanism, in the Druid revival, and in Nature spirituality.

In the Earth Mysteries movement, too: in 1969, the year of the Moon landing, John Michell published *The New View Over Atlantis*, an exploration of ley lines and secret geometry. The notion of ley lines was first introduced by Alfred Watkins in 1925, who noticed in the landscape an uncanny alignment of prehistoric sites and natural landmarks, which he theorised followed lines of energy criss-crossing the Earth. Watkins was hardly taken seriously at all, if not soundly ridiculed, for his proposition, but in the syncretic cultural context of the Sixties his ideas gained ground and developed into the Earth Mysteries movement, of which Michell was a major figure. The Dragon lines of Feng Shui practitioners, the Nazca alignments in Peru, or the song lines of the Australian Aborigines seemed to confirm that there was more to the ley lines concept than the mere delusions of a bored Englishman. Michell relentlessly explored these connections, finding in ancient geography and in prehistoric archaeology unmissable clues that not only were the Ancients perfectly aware of these telluric currents, they actively and purposely worked with them in a very conscious way:

> The prehistoric alchemists were dealing with the earth itself, which they regarded as the retort for the alchemical fusion between the 'sulphur' of solar or cosmic energies and the 'mercury' of the earth spirit. From this fusion were generated the forms of energy on which the prehistoric civilization depended.[4]

Suddenly the Earth appeared like a giant grid, a unified energetic field, connected through countless energy lines – resembling the acupuncture lines criss-crossing the human body. Many creation myths were uncovered by these modern terranauts that seemed to confirm the view of Earth as a giant web. Great circles were seen around the Earth englobing iconic sacred natural landmarks like Uluru in Australia or Palenque in Mexico. In the 1980s Robert Coon further proposed that two great dragon lines were encircling the Earth, one male, one female, meeting at two places, Lake Titicaca and Bali. Borrowing from Aboriginal mythology he referred to them as the 'Rainbow Serpent'. Coon also produced a planetary chakra map, with the first chakra, the root, located at Mount Shasta in California, and the seventh one, the crown, at

Mount Kailash in Tibet. The 'Earth Mysteries' movement is still very much alive today but, as can be expected, it is not taken seriously by mechanistic science and is easily dismissed as New Age wishy-washy nonsense. Yet it does offer another angle into a holistic and unified vision of Earth. It enlivens the beautiful planet that is our home. It also questions the contemporary 'Myth of Progress' that barely conceals its contempt for our ancestors and their supposedly crude technology. When taken seriously it gives a healthy dose of humility, both in our attitude to Earth and to our human lineage.

But the Waste Land – how heartbreaking it is to acknowledge: this is what planet Earth has become. Desecrated, raped, debased, commodified, exploited, abused. An 'It' to be subjugated at will. The Fisher Kings of this world, bleeding in the groin, are neither able nor willing to nourish the land. Their only concern is to grab what is left of it before all is gone. Are we running out of metals and minerals? Instead of making sure that we properly close the loop of consumption, we marvel at our endless ingenuity. Enter deep-sea mining – scraping the bottom of the oceans to recover the last nuggets, destroying all life down below out of sheer carelessness and mindlessness. And when we are done with Earth, having squeezed the last drop of juice, we will move on to mine the Moon. In the process, the Waste Land has become the Land of Waste. Mountains of it – in the earth, in the oceans, in the air. In technology we trust – anything that allows us to shirk a hard, deep look at ourselves. "All problems of existence are essentially problems of harmony," said Aurobindo,

> they arise from the perception of an unresolved discord and the instinct of an undiscovered agreement or unity. To rest content with an unsolved discord is possible for the practical and more animal part of man, but impossible for his fully awakened mind, and usually even his practical parts only escape from the general necessity by either shutting out the

problem or by accepting a rough, utilitarian and unillumined compromise.[5]

All of us, I dare say, long for unity. It is the unintegrated parts of ourselves that get in the way. Then we suffer. The Earth suffers. The community of life that sustains us suffers. This might sound like a cliché but a terrible sorrow sometimes overwhelms me when I realize, with an acuteness that is usually obscured by daily life, the poignancy of the Earth and the suffering that we inflict on animals, plants, and rocks. When I realize that I, too, as much as I would like to be a Grail Knight, am in many ways a Fisher King.

In the last few years, meat consumption has become a hot topic. From an environmental perspective, it is wreaking havoc. For many people it is also an ethical imperative to stay away from meat, given the terrible conditions in which animals are raised and killed. Yet we are now much more sensible to the plight of animals than we used to be; we recognise that they are sentient beings and that they do experience suffering just like humans do. We have come a long way from Descartes' indifference to the screams of the dogs he was dissecting, dismissing them as the mere "creaking of a machine". But 65 billion animals, growing up in appalling conditions, are slaughtered every year to satisfy our carnivorous drives. Chlorine-washed chickens, really? This is a huge shadow on humanity, and when we will have the courage and wisdom to face it, the collective grief will be unbearable. But where does the chain of suffering stop? "Every time I walk on grass I feel sorry because I know the grass is screaming at me,"[6] said McClintock. If grass is screaming when walked over, will we one day become aware of plant suffering and stay away from eating plants too? For German philosopher and philanthropist Albert Schweitzer, the ethical stakes are high:

A man is truly ethical only when he obeys the compulsion

to help all life he is able to assist, and shrinks from injuring anything that lives. He does not ask how far this or that life deserves one's sympathy as being valuable, nor, beyond that, whether and to what degree it is capable of feeling. Life as such is sacred to him. He tears no leaf from a tree, plucks no flower, and takes care to crush no insect.[7]

The reverence for life that Schweitzer was advocating allowed no compromise. Compassion cannot be ring-fenced; it is all-encompassing. Yet Schweitzer also recognised the unavoidable conflict at the heart of life: that in order to live, we have to kill. "In a thousand ways," he said,

> my existence stands in conflict with that of others. The necessity to destroy and to injure life is imposed on me... I become a persecutor of the little mouse which inhabits my house, a murderer of the insect which wants to have its nest there, a mass-murderer of the bacteria which may endanger my life. I get my food by destroying plants and animals.[8]

I wonder: if a lion had our consciousness, would it stop eating antelopes? Or would it breed antelopes by the thousands in packed farms so it does not have to worry about the hunt any longer? I am not a vegetarian but I do not eat much meat and I stay away from the industrial, non-organic sort. Am I shutting out the problem? Accepting an unillumined compromise? I accept that I may be – but I sometimes feel that vegetarianism may be cutting corners, refusing to acknowledge the inevitable: that is, that death is part of life. But not all deaths are alike. Dignity, gratitude, respect, thanks for the animal that just gave its life: indigenous communities knew all about the art of propitiation. The hunt was a sacred endeavour. We just kill, brutally, casually, flippantly. Soulless murderers. A choice that is entirely of our own making and for which, therefore, we have to bear full responsibility: "in ethical conflicts," said Schweitzer,

> man can only arrive at subjective decisions. No one can decide for him at what point, on each occasion, lies the extreme limit of possibility for his persistence in the preservation and furtherance of life. He alone has to judge this issue, by letting himself be guided by a feeling of the highest possible responsibility towards other life. We must never let ourselves be blunted. *We are living in truth, when we experience these conflicts more profoundly.* The good conscience is an invention of the devil (my italics).[9]

Paralleling the *Apollo* return journey back to Earth, I too have felt called back to Her. This longing, this call, spurred me to explore Nature spiritualities and led to my involvement with the Gaia Foundation. Looking at my birth chart, one may say that Earth had been there all along. In astrology, Earth is usually placed at the opposite end of the Sun. In my own horoscope this means that Earth is conjunct the Moon. Earth and Moon are locked together in a tight embrace, and they both face the Sun. Put another way, my Full Moon shines a bright light on Earth. I love the symbolism. Besides, Sagittarius is the sign of my Ascendant, which, in conventional astrology, makes Jupiter the 'ruler'[10] of my horoscope. In esoteric astrology,[11] however, Sagittarius is ruled by Earth – and so, my esoteric ruler is Earth.

Much of the work that the Gaia Foundation is doing is informed by the teachings of Thomas Berry, a true elder, an Earthling who, like McClintock, had a feeling for the organism. Berry was advocating for a true partnership with the Earth Community based on what he called an Earth Jurisprudence (EJ). EJ recognises that the primary law is the law of the Earth. All the rest, in a complete reversal of our usual myopic and anthropocentric vision, is secondary to it:

In our discussion of sacred community, we need to understand that in all our activities the Earth is primary, the human is derivative. The Earth is our primary community. Indeed, all particular modes of Earthly being exist by virtue of their role within this community. Failing to recognise this basis relationship, industrial society seeks to subordinate the entire Earth to its own concerns, with little regard for the consequences for the integrity of the planet.[12]

Borrowing the image from the alchemists, Berry was talking of the Great Work that had to be accomplished: the imperative to establish a new relationship with Earth, one that demanded "a largeness of vision and a supreme dedication". The *lead* of objects had to be transmuted into the *gold* of subjects: "as we recover our awareness of the universe as a communion of subjects, a new interior experience awakens within the human. The barriers disappear. An enlargement of soul takes place."[13]

For indeed it is through soul that we will be able to re-engage with Earth. Ecotherapist Bill Plotkin talks about Soulcraft, the tender care and attention given to the soul. By spending time in Nature, we encounter the *Anima Mundi*, the World Soul, and our own soul suddenly expands ecstatically, a giant sphere inflated with cosmic breath. We do not belong any more to ourselves only – what a relief! We commune with the Soul of the Earth, and we are changed by it because Nature acts as a Mirror:

In time, we encounter reflections of our deepest individual natures and perhaps hear our true name spoken for the first time. We come to understand that what is reflected by nature is not just who we are now but also who we could become. And so we begin entering nature as a pilgrim in search of his true home, a wanderer with an intimation of communion, a solitary with a suspicion of salvation.[14]

Bill Plotkin is a wilderness guide, leading nature-based soul-initiation ceremonies and rituals, many of them directly drawing on indigenous wisdom and traditions. The Vision Quest is one such ceremony and has been performed for millennia in indigenous cultures. At times of transition in a person's life, especially at the onset of puberty, people are being led in Nature, alone, vulnerable, afraid, with no food or water, no protection, and they have to face the Great Mother for days, sometimes weeks, often in life-or-death situations. This is a time of intense confrontation with Nature and, therefore, with oneself. Teenagers are meant to find their secret name – their true individual core, their purpose on the Earth, as individuals and as members of the community. Nature is the teacher.

In 2018, I participated in a Vision Quest in Wales under the guidance of an English woman, an elder of great stamina and compassion. I spent four days alone on the mountain, fasting, with a tent and duvet, no books, no phone, no music – just a drum and a notebook. Time takes on a different hue when one does not have the ritual of meals, the reckoning of time, the abundance of distractions. Fasting has the great benefit of clarifying and sharpening the mind, of uncluttering the chatterbox. The unconscious is vivified too: my dream life was extraordinarily intense, and it often felt that daytime had this dreamlike quality too. On the final night, I had to perform a small fire ceremony, holding a clear intention in mind. I wanted to ask for forgiveness and atone for the break-up of my marriage. The rest of the night was going to be a Vigil – staying awake till dawn. The weather had been appalling ever since I had set foot on the mountain, and rain was a strong possibility. When evening came, I pledged to stay outside and not cower back to my tent, come rain or shine. I had set up my own sacred space, my *temenos*, a mini version of a stone circle, with a small fire pit in the middle. The early evening was beautiful. I could see Orion rising in the East and I performed my fire ceremony, home and dry. But later on

– it may have been around eleven, although I had no clock to confirm – the weather shifted. It started to rain, gently at first but then with increasing intensity. I was wearing waterproof gear but the wind was relentless and it was cold. I held on but there is no denying that it was challenging – and yet, I had it so easy compared to so many indigenous people over millennia, who did not have the luxury of Gore-Tex! And I only had to endure one single night... After a while I lay down on the grass in a foetal position. There I was, a baby in the hold of the stone circle, alone and hungry, battered by the elements, an Adam expelled from Eden. I felt Her power resounding through my bones, shaking up every fibre of my being. There was something deeply archetypal in that night. When dawn finally came, I was reborn, emerging from the darkness of the womb, grateful for the new day, washed and rinsed and cleansed as never before.

The night following the Vigil, when I was back in the cottage and warm in my bed, I had a dream of terrifying intensity. I give titles to my big dreams, and this one I called *The End of the World* dream.

I am in a kind of giant glass dome and it is night-time. The world is very hot but nobody seems to be paying much attention. Suddenly we know that a huge deflagration is coming, a giant water bubble of extraordinary power. I brace for the impact. I am dragged under water and I am afraid that objects and debris will hit me, but in fact I am all alone in the ocean. I resurface. When I do, the world is devastated and TV channels are running the doomsday story, commenting that what had been predicted by many voices has happened. These voices had been silenced, or ignored, or mocked – even tortured. I am then taken into one of these torture chambers to witness what had happened. It is horrible. There is blood on the walls, and two arms are dangling from chains fixed to the wall. I can see to my right a man with an expression of extreme pain on his face. A sword had been thrust through the back of his neck, going all the way down to his right foot. I am now back watching the TV news, which claims that only one place in the world has been spared destruction, and that this place is located in Tokyo. A TV helicopter arrives on the scene, hovering over a high-rise building, a hotel. It is night-time and the whole atmosphere of the city is reminiscent of Blade Runner. *High up on the last floors the helicopter spots a swimming pool, like a giant pod. Many people are there, and they are waist deep in the water. They are all wearing tuxedos and evening dresses. They are aware of the catastrophe but continue to enjoy themselves. The atmosphere is absolutely decadent, yet creative, reminiscent of a Cabaret in Berlin in the 1920s. It feels like there is no respect for anything, it is all about complete hedonism. People hold champagne glasses and wave at the TV cameras. On a little stage by the pool, two violinists are about to play. They have been taught to perform a little choreography, like trained monkeys. Nothing is sacred in this place, everything is turned into an object of fun and amusement with complete cynicism. Then I am back to the place where I surfaced from under water. There are survivors and I recognise a few faces that I associate with the Gaia Foundation. Someone asks: "So what do we do now, should we split?" I seem to be the one in charge and I say: "Not at all, on the contrary we must stay together, we will be stronger." A circle of people is spontaneously forming. In the centre are women with babies and toddlers, on the outer rim are men who will have to defend the circle. I realize that we are now in complete chaos and that nothing is safe anymore. I say to everyone: "The sad reality is that from now on, anyone approaching us has to be considered as a potential enemy."*

This was a highly disturbing dream. The atmosphere was one of total annihilation. Not only was the physical world destroyed but it felt like an obliteration of Spirit too. Maybe that is what Hell looks like. What was left of humanity was either totally cynical, or absolutely terrified. The Earth was ravaged. Coming up immediately after the last night of my Quest, I somehow felt that it was Gaia herself who had sent me this haunting dream.

The power of the Vision Quest resides in this willingness to expose oneself to the terrifying embrace of the Mother, to offer oneself up sacrificially, small and weak and vulnerable, to bow down with humility in front of our Ancestors, the long lineage of which we are the latest avatar – all of our Ancestors: our human ancestors; our animal ancestors, our forebears of skin, claws and feathers; and further back, rocks, trees, rivers, clouds; the Earth herself, our cradle.

Deep in the bosom of the Earth, I was the one and the many. "Now I see the secret of the making of the best persons," said Walt Whitman. "It is to grow in the open air, and to eat and sleep with the earth." And so I, too, echoing the great American bard, sing the song of myself.

I am the 40 million-year-old whale, screeching and bellowing in the amniotic fluid of Mother Earth.

I am the soil, remembering how dragons used to nest in blazing cauldrons.

I am the tiger, blood of the forest, bruised and beaten, seeking refuge in the warm embrace of silk.

I am the stream cascading on the rocks, which knows the price of beginnings.

I am the bark of the beech, whose elephant skin gleams on the neck of the forest.

I am the stone circle that knows no bounds, caught in the desire of a new-born eagle.

I am the testicles of a bull, exchanging fluids with the setting sun.

I am the shade, hiding in the bosom of a decaying root.

I am a smile, ripped from the heart of a feasting lion.

I am the tower, cracked open by the sound of copulating ants.

I am the raven of the morning star, twisting and turning and flying high into *my eyes*.

I am the green smoke of the forest, wiping out the memories of countless dwarves spitting blood in the bowels of a dying city.

I am Bear, trapped in the jaws of pitiless hunters.

I am the daffodil, whose eye pierces the inside of my soul.

I am a glowing crystal of consciousness.

I am That.

Tat vat asim.

I spent the year 1987 in Saudi Arabia, working as a civil servant for a short-term posting. I was 23 and, maybe surprisingly for such a dry land, I took up scuba diving. After completing the proper training there came the day to dive on my own, without the protective guidance of an instructor. I went with a friend, a young fellow like me with hardly a few dives under his (lead) weight. We were in the Red Sea, near Jeddah, and we had been told of a good spot where we could find a shipwreck about 15 metres deep. Shipwrecks are much appreciated by divers for they are usually brimming with wildlife. We went down and the spectacle was magnificent. There were indeed many fish in the vicinity of the wreck, including a majestic *murene* [eel]. But suddenly we spotted a shark coming our way. I remember to this day the absolute terror that overwhelmed me – a terror that expressed itself in these very simple terms: "You have nothing to do here. You are a stranger in this land and this shark has every right to do as it pleases with you." Yes, it was this feeling of being totally alien to this strange underwater land that hit me immediately. The feeling of having stepped into a no-go zone, of being an intruder, of having violated a boundary. I was not welcome there. The shark was probably no more than a small reef shark and any seasoned diver would have quite happily played with him. We did not. After exchanging pathetic distress signals, we promptly left the wreck and made our way back to the surface. The whole dive lasted no more than fifteen minutes, yet my oxygen bottle was empty. It was meant to last 45 minutes: the heavy breathing occasioned by the sight of the shark had sucked the air at frightening speed. I dread to think what would have happened if we had met the shark later in the dive: I would have run out of gas in the middle of my heightened panicky state. But here is not the point. The reason for my telling the story is to reflect on this overwhelming sensation of being an alien in wilderness. Of being alien *to* wilderness. I did not belong to the deep ocean. I was a city boy. Nature was to be feared. I had left the protective embrace of civilization and I was going to pay dearly for this foolish transgression.

Of course, an additional reading, a symbolic one, is possible. Just like my meeting with the three snakes symbolised the transformation I was going through, I could see in retrospect that my terrible scare of the shark may have symbolised a confrontation with an unconscious content that terrified me. At the time I was not looking at the world with symbolic glasses and, thirty years later, I cannot remember precisely what I was going through. All I can say today is that the emotional intensity of that encounter may in all likelihood have mirrored a deep inner process.

Three years ago in South Africa, another perspective was offered to me. As our group was about to depart for our Vigil, our night awake on the mountain – alone and with no fire or protection from local wildlife – one of the medicine men said a few quiet words. You have every right to be anywhere you want on this planet, he said, because you are an integral part of it. You are going to spend the night in nature and it is your God(dess)-given right to do so. You are an animal sharing the space with all the other animals. This creates rights and obligations but above all, it means that you are not unwelcome strangers. It also implies that you are vulnerable, as all animals are, and you have to embrace this vulnerability. It is the gift and the teacher.

I went through the night with some trepidation and anxiety. We had to carefully choose our spot for we had to stay put for twelve hours: walking around in the dark of the night is not a good idea. I found a little grassy platform with a flattish stone behind me to support my back. Perfect, a wonderful observation spot. Now I could enjoy the spectacle. The weather was glorious but very windy and cold. The sun set quickly. For a short while the whole landscape became almost frantic: amazing to witness the activity taking place at sunset! Baboons could be heard barking in the distance – mothers gathering their unruly offspring, I surmised. I could see all sorts of insects flying by me, no doubt going back to their nests or refuges after a day of foraging. Commuters of the wild! As darkness enveloped me, Venus became visible high up, bright and shiny and slowly – oh! so slowly – she dived towards the horizon. It took her several hours to set. Before disappearing she turned red, almost Martian-like. Then shooting stars came raining down, like a firework. I must have counted twenty at the very least. It was bitterly cold, and I sang a few French songs to warm me up. I wondered if French had ever been spoken in this wild place. The Moon rose in the middle of the night, majestic, blinding the stars gathered around her like courtesans. At times it felt like the night would never end but suddenly I heard a cockerel somewhere far below. Was that it? Was dawn pointing? As I was pondering this, I distinctly heard a growl and I froze. For a few anxious minutes I was very guarded, for I knew that leopards were occasionally roaming the area. But leopards are very shy creatures: they will carefully, and wisely, avoid human contact. My leopard vanished into the last remnants of the night. It was the only time of the night when I really experienced fear. The sun rose at last. A new day.

To offer one's vulnerability to Nature has a mysterious affirming effect. It is extraordinarily liberating. One feels renewed, brimming with inner strength and steadied by a calm purpose. In shedding one's carapace, one learns to *trust*. The body is bristling with a delightful tingling sensation – life coursing through one's veins – while the soul is soothed by an overwhelming gratitude for being alive. By shedding the subtle shackles of civilization, by making peace with the *fear* of the wild, one sings. One crows. One snorts and growls. One chirps and snarls and squeaks and hisses. One is.

"Whenever my life devotes itself in any way to life," said Schweitzer, "my finite will-to-live experiences union with the infinite will in which all life is one, and I enjoy a feeling of refreshment which prevents me from pining away in the desert of life."[15]

For ecophilosopher David Abram, the rupture of our dialogue with Earth had a lot to do with the invention of writing: "it is only when a culture shifts its participation to [the] printed letters that the stones fall silent. Only as our senses transfer their animating magic to the written word do the trees become mute, the other animals dumb."[16] We have lost the ability to communicate with Nature, ensconced in a human world that is self-referent. The loss of this great conversation, said Thomas Berry, has degenerated into 'spiritual autism'. Our great crime, somewhere along the way, has been to lose our reverence for Earth, "to lose," as Zen teacher Susan Murphy points out, "the humble amazement of feeling the Earth and ourselves to be a prodigious unearned gift from the universe." "How," she asks, "did we make ourselves self-appointed lords of the Earth and, like spoilt children, throw all gratitude and humility out the window?"[17] For environmentalist Paul Shepard, the consequences of this "historical march away from nature" can be seen in "massive therapy, escapism, intoxicants, narcotics, fits of destruction and rage, enormous grief, subordination to hierarchies that exhibit [a] callow ineptitude at every level, and, perhaps worst of all, a readiness to strike back at a natural world that we dimly perceive as having failed us."[18]

The advent of digital communications has vastly contributed to accelerate our disconnection: "caught up in a mass of abstractions, our attention hypnotized by a host of man-made technologies that only reflect back us to ourselves, it is all too easy to forget our carnal inheritance in a more-than-human matrix of sensations and sensibilities."[19] We are in a dialogue with ourselves only, and, eyes glued to the screen, we are oblivious to the tragedy unfolding around us. Earth has become e-arth: a virtual planet, one with whom we rarely engage any more – and when we do so, by donning digital gloves. We have, literally, put a screen between Her and us, and we are increasingly estranged:

> Transfixed by our technologies, we short-circuit the sensorial reciprocity between our breathing bodies and the bodily terrain. Human awareness folds in upon itself, and the senses – once the crucial site of our engagement with the wild and animate earth – become mere adjuncts of an isolate and abstract mind bent on overcoming an organic reality that now seems disturbingly aloof and arbitrary.[20]

Abram pleads for the rediscovery of our sensual being, for the reclaiming of our bodies, for a re-engagement with the natural world – the more-than-human world, as he calls it – through the senses. It is a daunting task. "Civilized man… is in danger of losing all contact with the world of instinct," said Jung, "a danger that is still further increased by his living an urban existence in what seems to be a purely manmade environment. This loss of instinct is largely responsible for the pathological condition of contemporary culture."[21] At a deeper level, Abram thinks that many people are simply overwhelmed by the physical suffering experienced by planet Earth: the dissolution of ecosystems, the loss of species, the pollution of air and water – "one of the penalties of ecological education is that one lives alone in a world of wounds," as Aldo Leopold used to say. "I encounter many people who are frightened of their direct, animal experience," says David Abram, "who are terrified at the mere thought of trusting their senses, and of stepping into a more full-bodied way of knowing and feeling, because they intuit that a more embodied and sensorial form of awareness would entail waking up to so many grievous losses."[22] Screen time offers an escape route: "They want to retreat into relation with their smartphone or their iPad, taking refuge in the new technologies with their virtual pleasures. Because they quite rightly sense that there is some grief lurking on the other side of such a corporeal awakening."[23] But grief, he says, echoing Joanna Macy,[24] is a "necessary threshold through which each one of us needs to step." "It's as though the parched soil underfoot needs the water of our tears for new life to begin to grow again."[25] *We are living in truth, when we experience these conflicts more profoundly.*[26]

I once attended a week-long workshop in Schumacher College, Devon, with David Abram. I do not think that I ever met a human being so deeply attuned to the Earth. At the time of the workshop, raging fires were destroying vast tracts of forests in the West of the United States, near his home. When David was talking about it, it felt as if it was his own flesh that was on fire.

Abram does not have a romanticized, airy-fairy notion of life on Earth. He acknowledges the power of the Mother and the dangers and fears that humans have to face. The Earth is not benign. The Earth is not gentle. The Earth is fierce. But with the recognition and willing acceptance of our vulnerability comes an extraordinary boon:

The palpable world, this blooming, buzzing, wild proliferation of shapes and forms that feed upon one another, yes, and yet also jive and dance with one another – this earthly cosmos that our work is trying to coax people into noticing – is not a particularly nice world. It's not a sweet world. It's shot through with shadows and predation and risk... but it's mighty beautiful, it's shudderingly beautiful precisely *because* it's so shadowed and riven with difficulty.[27]

"When we began to seek protection from the life-threatening

dangers of the natural world," said ecologist Claire Thompson, "we were trying to avoid death. Now, however, by locking ourselves up indoors and in our minds, it seems we are beginning to avoid life."[28]

One must accept to live with dirty broken nails.

I must accept to live with dirty broken nails.

Epilogue

Part 1

By focusing so much on the interplay of Sun and Moon, am I stuck in duality?

Lord Siva has compassionately revealed dualistic scriptures for the benefit of those who see reality that way; and even in these texts, He has concealed hints and clues that point toward nondual awareness so that those in the dualist path may, when they are ready, 'graduate to a more inclusive nondual awareness'. And lest they then become stuck in nondual rejection of dualistic modes, there is the teaching about the 'higher nonduality' that includes all possible modes as the levels of emanation of the One.[1]

The Sun, the Moon, the Earth – they all lead back to the Source. Pathways to ultimate knowledge, like you and I and the ten thousand things.

Part 2

I

I have come a long way. At the beginning of the year 2019, as I was working on this book, I decided to visit my daughter who had been living in Mexico City for two and a half years. Today, one cannot consider flying without an acute feeling of guilt. Moreover, how can one write about Gaia and board a plane? There is an inherent contradiction, an impossible squaring of the circle. But we live in a time of transition and one must accept the tension. Most environmentalists criss-cross the planet to spread their message. Are they mere hypocrites? I think one must develop as much consciousness as possible. Maybe in some circumstances flying, despite its inevitable destructive trail, has a benefit that more than compensates for it. I know how flimsy this sounds. We face the very same conundrum in our everyday life: what we eat, how we eat, what we wear, how we travel, how we keep warm. I read once about an English woman whose daughter was living in New Zealand. This woman had pledged never to board a plane again and, since her daughter was financially unable to fly back, she had basically accepted that she would never see her again. That is a very high price to pay. In the context of my work, the timing felt right for this trip: I was eager to explore the compelling power of the Sun and the Moon on pre-Hispanic cultures, and I had an overwhelming impression that an answer was awaiting across the ocean. In a sense, I was killing two birds with one stone – with due apologies for the gross image. After Mexico, I felt powerfully drawn to go back to Titicaca. Ten years ago, on the Peruvian side, I had had a deep realization so it seemed appropriate to go back. But this time I would go on the Bolivian side to visit the Islands of the Sun and of the Moon. The Island of the Sun was said to be the birthplace of Wirakocha, the powerful Creator god of the Incas who was also worshipped as god of the Sun. The Island of the Moon was home to the temple of the Virgins of the Sun. Both places were said to exude a mysterious energy, and the lake itself, the sacral chakra of the Earth according to Robert Coon, is often held to be the meeting place of Heaven and Earth – of Sun and Moon.

II

I was standing at the bottom of the stairs of the Temple of the Sun in Palenque, Mexico. The Sun of the Palenque Mayans was a midnight Sun. They called it the Jaguar Sun. The jaguar is a nocturnal animal, associated with lunar power. In Palenque, the Sun was worshipped for its ability to fight and defeat the forces of darkness and to emerge victorious at dawn. As I was about to climb the stairs I mechanically looked up. A middle-aged

Mexican woman was standing at the top, all dressed in white. She was all smiles and standing in the characteristic pose that one adopts for the ubiquitous selfie, but what struck me was the motif embroidered on her blouse: a big yellow Moon was pictured alongside a bright red Sun. There it was, the sacred marriage atop the pyramid of the midnight Sun.

III

I arrived in Copacabana, on the Bolivian side of the Titicaca lake, just as the Moon was about to enter Leo, the sign of the Sun. My initial plan was to stay for a few nights on the Island of the Sun and maybe stay overnight on the much smaller Island of the Moon. But I had been warned: something was amiss on the Isla del Sol. The island was cut in half, between the Southern and the Northern sides. The South was open to tourism while the communities in the North were refusing to allow foreign visitors. A few days before my arrival, a community leader of the island had been kidnapped and imprisoned by the Bolivian authorities. Rumours abounded of a disturbing ritual killing a few months before. Before making any further plans I decided to visit both islands in turn. As it turned out I went to the Sun on Sunday the 12th and to the Moon on Monday the 13th. Symbolism is sometimes wonderfully accurate. The contrast between the two islands could not have been greater, though. The Southern side of the Sun was indeed brimming with tourists and the inhabitants seemed somewhat nervous, rattled by too much tourism and too overtly eager to cash in on it. The overall feeling on the Moon was totally different. Hardly any tourist was staying overnight and the island exuded soft and quiet. The community seemed at peace with itself and cleverly keeping a tight rein on tourism. This felt like a very appropriate metaphor for our times. Solar consciousness was suffering from overkill, Marion Woodman had said. This was perfectly embodied in the fractured and unsettled isle of the Sun. What it desperately needed was to recover the apparent harmony of the isle of the Moon, a mere seven kilometres away. I then decided to change my plans: I would stay eight days on the Island of the Moon, the length of time of the *Apollo 11* venture. I would lodge in a modest hostel run by a lovely couple. There I would put the final touch to the book. Besides, right in the middle of my stay, at the equivalent symbolic time of the *Apollo* Moon landing, the Moon in Scorpio would be Full. A fitting conclusion.

IV

Day 1: Before reaching the Isla de la Luna, one stops on to the Isla del Sol, which houses the temple of the Sun, dedicated to Wirakocha. Nested in one of the several rooms of the temple, a giant stone is said to represent the heart of the God. As the fourth of the seven chakras, the heart reconciles the lower three chakras with the upper three. It is where earth and heaven energies meet and blend.

While walking on the island I spotted an eagle overhead, circling into the blazing sun. How neat a polarity would it be if I could then meet a snake on the Island of the Moon! Much smaller than her counterpart, the Island of the Moon is an elongated rocky outcrop hosting no more than 25 families. On the north-western side, one can admire the remains of the Temple of the Virgins of the Sun, displaying an abundance of Andean crosses. The Andean cross symbolises the four quarters of the yearly course of the Sun as well as the four midpoints – mirroring the Celtic Wheel of the year, with its eight spokes. The temple has a somewhat chequered history. The Virgins of the Sun, it is said, were young girls groomed for the Inca nobility. They were essentially weavers of clothes – weavers like the Moon – and were chosen for their beauty. Once married to high-ranking Incas, they would leave the Moon to go to the Sun. Was the temple a prison, a symbol of the abuse of the feminine? Or did it house a feminine mystery school as well? What is beyond doubt is that as soon as the Spanish invaders set foot on the island, the Virgins of the Sun were brutally sold as slaves.

Day 2: The Island of the Moon IS the serpent. A local legend relates how a giant snake was swimming towards the isle of the Sun when an Inca slaughtered it before it could land. The body of the serpent, turned into stones, became the isle of the Moon. The north-western part of the island is referred to as the head of the serpent, and the south-eastern one as the tail. Another serpent-slaying myth, then. There seems to be no end to those. In Hindu astrology, the serpent's head and tail refer to the nodes of the Moon, respectively the North Node and the South Node, Rahu and Ketu. Vedic astrology is very fatalistic and Ketu is regarded as potently malefic for it symbolises past behaviours that have to be let go of. Ketu is the place where the serpent excretes and one should not linger there.

As the day ends, I make my way towards the head to watch the sun set. There is absolute tranquillity and stillness in the land. Llamas are silently grazing, throwing quizzical looks as I walk by. Standing atop the red cliffs, I witness the declining course of the sun over the lake. To my right, up above, the Moon, not yet full, has risen. I hold Sun, Moon and Earth in my awareness. At the altitude of the lake, close to 4,000m high, the colours are exacerbated. So is the silent music of the spheres. I reflect on the long journey that brought me to this point of equilibrium, where the earth seems to touch the sky.

Day 3: Exactly ten days ago…

All of a sudden, a huge wave is coursing through me. I can feel the plant taking over, reaching to the last cell of my being. My body heat increases rapidly, I desperately want to rip my clothes off. The overall feeling is very uncomfortable. I am rocking back and forth, trying to shake off the intrusion – anything to get me out of my skin – while simultaneously surrendering to it. The visions start immediately, image after image after image, hundreds of them at breakneck speed, a merry-go-round of colours and shapes. A short while later, the unpleasant feeling in my body has subsided and I give in to the fireworks. Suddenly, in the corner of my left eye, a child appears and seems to beg me forward. I follow him, down and down, until I notice the faint contours of a house. Approaching ever closer, I am hovering over it and I know that it is the house in Brittany, France, where my parents took us, my brother and I, on holidays a long time ago. I can see the living room, in which a little boy – me, 18 months old – is sitting with his mother. Suddenly she rushes out, screaming, and the little boy is left all alone, unable to comprehend why she left him, a terrifying sense of dread enveloping him. The feeling of abandonment, vulnerability and fear is unbearable. At this precise moment the boy takes a radical decision. Never again will he allow himself to feel that way. He makes a pact with himself and a split occurs. The boy that came to earth is left for a different boy. In my wild journeying to the depths of my own being, an image imposes itself: I can see the boy encased in a kind of steel tube, with a steel armour around him. Suddenly it occurs to me with absolute certainty that my whole life has revolved around this point in time, this knot, that everything that followed found its origin there. I see a huge deflagration expanding from this moment in time and all the events of my life unfold. I am overwhelmed by a feeling of utter tenderness for this young boy who did not know how to cope. I forgive him, even though I know that the pact he entered into was a betrayal of soul that took 54 years to unravel.

I had meant to meet Grand Mother Ayahuasca for ten years – ever since I heard about her in Peru. But like a young girl who wants to keep her virginity until she meets the right man, I wanted to wait for the right conditions to meet Her. The stars had to be aligned. Set and setting, Timothy Leary used to say – something like that. Encounters with plant spirit medicine cannot be taken lightly. The plants are not recreational drugs. They are sacred beings. One has to prepare and be ready. One has to acknowledge the reverence of the teaching. While I was about to ingest the sacred brew, in a kiva outside La Paz charged with potency, and under the guidance of a young shaman whom I felt I could entrust with my soul – because this is what a medicine journey is, and one should be fully aware of it – I knew the time had come.

When I was 18 months old, my father had an accident. While holidaying in Brittany, the fence against which he was leaning gave way and he fell down six metres below on his back. My brother, two months old, myself and my mother were in the house when it happened. I have no conscious recollection of the event, and the trauma that my parents went through put an iron lid on the event. My father would never talk about it, and my mother hardly ever. Of course, I understand why. It took me decades, however, to realize that this event may have fashioned me in some way. What the Grand Mother showed me with terrifying clarity was that it was the cornerstone of my life. I had betrayed myself, She unequivocally said.

When I was nine or ten, I used to play a little game. I would stand in front of the mirror and look at myself straight in the eyes. A minute or so later, a shift was inevitably occurring. I was sensing that somebody was hiding beneath the image of the boy, that *another me* was lurking behind. I would then run away, scared of what I had intuited.

It all made sense. I had always felt that I had created the conditions that put me in a French jail for three days. This dramatic episode of my life was a perfect replica of the little boy encased in steel, a mirror image that would shake me out of my slumber by the sheer force of resonance. In that moment the soul became aware of something deep that pointed back to herself. But it was only a first step, an intuitive and unconscious movement. The Moon had risen, however faintly.

Maybe personal and collective are meeting again in this story. I traded vulnerability for protection in the form of conformism, money and privileges. Out of fear, I rejected the power of the Mother for an illusion of security but by doing so I cut myself off from the spring of creativity and true individuality. This is the archetypal solar journey gone awry: the fashioning of an ego of steel – an ego that subjugates Nature, including one's own, an ego that, terrified of its own insecurities and vulnerability, hides behind an illusion of control and power. It is to be seen everywhere around us. It is in fact the dominant model, and it has been so for thousands of years. It is the scorching Sun. It is the Waste Land.

Day 4: It is early morning – dawn has hardly broken – and I am fully awake. I decide to get up and meet the Sun. I make my way to the tip of the island to watch sunrise. The course of the Sun in the southern hemisphere is always confusing for a Northerner, running at it does from right to left. I reach the tail of the serpent, a very rocky strip of land that tapers off into the lake. A double stone circle has been built by unknown hands, and it looks like it is oriented east-west. I enter the first circle through the West, facing East, and then the second, a much smaller circle only large enough for one person. Cross-legged, I can see the rays of the Sun darting across the lake behind the magnificent face of the Janko Uma, 6,388m above sea level. I am reminded of Apollo, wearing his crown of sunrays. On the lake a whole party of seagulls are quietly being rocked by the gentle swell. The halo around the Janko Uma is quickly getting brighter and suddenly there He is, a focused tiny dot at first expanding into a disc of heightened intensity. There is always something profoundly stirring to watch the sun rise. One feels expansive, filled with a strong desire to open one's arms and embrace the world. All of a sudden I hear a cacophony of noise around me. The seagulls are celebrating the return of the light in their own way. They have all taken to the air and delight in their shrieking. "Does the crow wonder if its cry is pleasant to the ear?" one of the Sangomas had said in South Africa. "It does not. It just is." I want to add my voice to the morning chorus of Creation, and I sing Kirtan, Hindu devotional songs. My voice is quite rough and unsettled in the early morning but gradually it clears up and I feel more confident to embrace the chant more fully. I feel profoundly at peace on this spot, in the embrace of stones, welcoming the return of the Sun with ancient holy songs that must have witnessed untold sunrises. Tears roll down my cheeks. Healing tears of heightened awareness. The Sun is individual. I, the scavenger.

At ten past five tonight, the Moon will be full. It will still be daylight then, the Moon will not be visible, but barely one hour later the Sun will set and the Moon will simultaneously rise in all Her glory. I want to meet her too. Walking on the beach towards the tail, I can see on my left the glowing halo of the Sun that has just set, while on my right is the bright rising Moon. One eye orange, one eye white indeed. I give thanks. Thanks for the grandmother medicine that shed such a bright Full Moon light on some dark corner of my psyche. I pray for further insights. I pray for us, children of the Sun and Moon – may we hear again the voice of the Lady of the Night, may we have the humility to rediscover her old and wise ways. With the Moon, one feels drawn inwardly. One folds upon oneself. One sings in silence the songs of countless generations gone by. One bows and prays and stands in awe, and in this inflowing reaches down to the depths of our human experience. No words necessary. The Moon is collective. We, the scavengers.

Day 5: By going to the Moon, we have gathered precious geological evidence. The Moon and the Earth are interlocked into a gravitational dance without which life on Earth, in all probability, would never have developed. One reason is that the Moon's pull stabilizes the Earth's axial tilt (currently 23.5 degrees), allowing the regular occurrence of seasons. Without it, the Earth would constantly rock back and forth on her axis, like a drunkard staggering in space. Furthermore,

It is highly probable that the extreme motion in the Earth's subsurface belts causing the continental plates to move around are a direct result of the gravitational attraction of the Moon very early in the Earth's history... Where continental slabs move around over geologic time, they create environmental niches where life thrives and is accelerated in adaptation because of the rapidly changing climate caused by plate tectonics.[2]

Therefore, "the Earth-Moon system has created a forcing-ground for the rapid evolution of life and it could be said that *we ourselves are only here because of the Moon*"[3] (my italics).

Day 6: Even more than Joseph Campbell, Carl Jung grappled with the notion of myth. For while Campbell looked at myths from the perspective of the collective, Jung also explored the individual myth. Today, a myth is dismissed as a mere untrue story. For the ancients, it was the opposite: a myth was a source of knowledge and truth. It was alive. When Jung broke up with Freud, he realized that he did not have a personal myth. He then descended into an extraordinary exploration of the psyche that he charted and recorded in the *Red Book*. After several years of this intense inward journey, he found his myth. Campbell's work was more outward in nature. Myths, he reckoned, performed four functions. Cosmologically, they told the story of the creation of the world. By so doing, the culture was connected to the mystery of its existence, to the *Mysterium Tremendum*. Collectively, they linked the tribe, or the cultural group, to that story – in particular through rituals – re-enforcing the sense of an orderly cosmos. Socially, they regulated the life of the community by morally bounding the members of the community to the group and to the wider story. Individually, they gave each and every member a blueprint for a life that, by being given context and meaning, could be lived with integrity.

Both Campbell and Jung were lamenting the contemporary lack of myth. What is the common story that binds us together? We have severed the connections both to the Heavens and to the

Earth, and we are left stranded in the middle with no sense of purpose, either collectively or individually. This is unbearable of course, so we turn to false idols. We marvel at the power of technology and fantasize that it will bring about the new human, more efficient, more clever, free of diseases and, hopefully, of death. We have been blinded by the light of the Sun. What indeed, one may ask, is the point of an eternal life lived without meaning?

Can the *Apollo* mission to the Moon be seen as a new myth? In the Greek story, Apollo slayed the Python, a child of Gaia. He henceforth took over rulership of Delphi, replacing the original ruler, Gaia. Seen with a symbolic lens, the 1969 lunar mission saw Apollo uniting with the Moon and this union *gave birth* to Gaia, a new vision of Earth. Traditional cultures often held myths to be the recording of literal events. Giants did create the Earth, Gods and Goddesses did battle against forces of darkness and evil, or among themselves, to bring about mankind. Rituals re-enacted these original events, enforcing the covenant between the group and these Ancestors. We do not need to adopt a literal view, in fact we cannot. However, at this juncture Imagination comes to the fore. By enabling an 'as if' perspective, we can be both in and out. Evidently, Apollo did not mate with the Moon on 20 July 1969; but we can engage with the whole venture as an 'as if' story that opens up new perspectives. The power of the symbol resides in its ability to point to something beyond, something that may not be real in a literal sense but that is very real for the soul. It is Blake's double vision. If we adopt this view, we realize that the true message of *Apollo 11* is not to wallow in our own ingenuity but to re-establish our long-lost covenant with Earth, the divine child of the sacred union of Sun and Moon.

Day 7: Like *The Odyssey*, like *Battlestar Galactica*, the myth of *Apollo 11* is a story of homecoming. We return to Earth with a new vision, a new way to engage. Owen Barfield identified three stages in the evolution of consciousness: Original Participation

(*participation mystique*, Great Mother), Separation (birth of the ego leading to modernity), Final Participation. We could roughly schematize in assigning lunar consciousness to the first stage and solar consciousness to the second. The work now is to move on to the third stage where we hold both lunar and solar perspectives. In a similar way, Paul Ricoeur described this third stage as Second Naivety. We cannot be naïve any longer, but we can adopt an 'as if' attitude that includes and transcends polarity.

On Titicaca lake, informed by the ancestral Andean culture, the power of the Madre Tierra, the Pachamama, is everywhere to be felt. On the road to the lake, one sign reads: "Take care of our sacred lake." This is something that we would not see in Europe. Sure, we too want a lake to be protected and kept free of pollution but it must be so for our own enjoyment, not because it is a sacred being. And yet, this is not to romanticize the Andes either. Some places are shockingly covered in rubbish. People just do not know what to do with plastics so they burn them, leaving the charred bodies of water bottles and washing-up liquids lying on the ground. But in the Andes one still encounters a very tangible reverence for the Earth, an acknowledgement that She comes first – something that we have lost sight of long ago in our cultures of Progress and Technology.

I believe that the *Apollo* journey shows the way forward. It unites Sun and Moon and it points back to Earth, the place where the sacred marriage is to take place. We are no longer unconsciously fused with the Earth, and we will never be able to return to that stage. It is not even advisable. But we can feel again that fusion through the power of the Imagination, a power that resides in the heart. The mind cannot make that move for essentially it is a move of compassion. What the mind can do is defer to the heart, not enslave it. It can acknowledge that it is in service and, why not, may even find delight in this service. In saying that, I know how long the path is for me, too.

Day 8: Gosh, how hot the Sun is! And yet one immediately shivers when retreating to the shade. On the Island of the Moon, ghosts were dancing the storm away. A whole gang of devilish faces with long pointy ears. It has been a long journey, to the Moon and back, but the road ahead is far longer. May we store wax for our candles! May we keep warm in the quivering lights of the cave! We are desperately searching for home and yet it is all around us. A smartphone is not a home. Neither is a rocket leaving Earth. Nightingale is, and birch and daffodil and snowflake and leaf of grass and crystal and feather and woman weaving and man carving and black Madonnas and angel wings and the love of candles and the jade-eyed snake and the rose window in Chartres and Isis unveiled and Orion reflected at Giza and the giant Pyramid of the Sun at Teotihuacan and the Dogons in rapture at sunset and a trumpet tearing apart the nascent dusk and a swine in the mud and Apollo driving his chariot and a soup of watercress and Viviana washing her sheets in the freezing lake and the eagle flying with the condor and blue orbs of light and primrose and the colours of a *chitanje* and clouds at breaking point and the wind on my face and the wind on your face, comrade, and the fragrance of musk and the healing power of lapis and the magi in the Age of Pisces and the orphans in Calcutta with jewels in their eyes and a budding twig and the ridge of a mountain and the women in Malawi toiling under the Sun and the spring waters of Mount Kailash and the glistening skin of a dolphin and the wild call of the Silk Road and the hairy legs of a spider and the Moon hiding in pools of water and the brutal metabolism of the jungle and a stupa with giant eyes and the foam of a wave and the one and a thousand nights and the Jaguar King crowned for all ages and the ice drip-drip-dripping in the fountain and the perpetual choir at Glastonbury and the glowing bark of the Tree of Life and the ecstasies of Teresa and Shiva-Shakti in Tantric union and Uluru bleeding ochre at sunset and the salt of the earth and *Wuthering Heights* in the bosom of a virgin and riders on the storm and light as particle or light as wave and Picasso hiding behind African masks and the magnetic gaze of Easter Island and Cundrie shaming me in my cell and the giant frame of the killer tree and the Sagrada Familia shedding tears and a journey down the rabbit hole and the Sirocco blinding the camel and Buddha touching the ground and a *mate de muña* and garlands of flowers round the neck of Ganesh and the humility of the pilgrim and shaved heads at Full Moon and dreadlocks at New Moon and the forlorn echo of Atlantis and the plight of the whale and the banter of market traders and the exquisite dance of the bee and the Mother opening her arms and the scent of spices and the shine of amethysts and a hole in the ground and the pulsating beat of the forest and Ramakrishna smiling an eternal smile and a freshly squeezed orange juice and the spark of Sirius in the winter sky and the howling monkeys greeting dawn and the sound of Taizé singing in the misty landscape of Burgundy and the Alpha spiralling with the Omega and the burning bite of *pimientos* and the Great Nebula exploding into billions of fragments and circles within circles of radiating light and the secret chambers of the soul and rituals at Lascaux and the pipe of Van Gogh and the bat patrolling the shade and the twists and turns of DNA and our Ancestors on the horizon and the smoke escaping from the hut and the bellowing of the stag and Dionysus on a pyre and dragon medicine and scales and nails and horns and our unfolding future receding in the distance and the Hermit leading the way and the smell of coffee and the Tao that cannot be told and invisible hands shaping the landscape and Mother Moon and Father Sun and the Divine Child – Our Blessed Earth.

One small step for Mankind, one giant leap for planet Earth.

Appendix

Astrology: When the Prince Kissed the Frog

Astrology is often described as the art of beginnings, grounded in the Seed Moment – the moment for which the birth chart is cast and which contains in itself the potential of growth that will unfold in time. Dane Rudhyar considered that the key astrological moment was the first in-breath at birth. When the umbilical cord is cut we disconnect from the mother and become independent units. We attune to the cosmos in which we are thrown, fill our lungs with cosmic breath, and the quality of that breath is reflected in the birth chart. This is the Seed Moment of one's life. Astrologers pay a lot of attention to the Seed Moment because for them, time has a quantitative *as well as* a qualitative nature. The Greeks recognised this dual nature of time: *Kronos* referred to quantitative time and *Kairos* to qualitative time, which carried the all-important notion of the right moment. Away from the traditional canons of modernity, where time is a quantifiable commodity, astrologers assign qualities to time and, in their vast majority, they consider that these qualities are *objective*. As such they see their role as that of an interpreter, an objective reader of an objective moment of time. This approach was rooted in Jung's initial work, who was writing in 1930 in his Memorial to Richard Wilhelm: "Whatever is born or done this moment of time, has the qualities of this moment of time."[1] To be as accurate as possible in the interpretation, therefore, astrologers must work with a precise time of birth. From then on, the potential latent in the chart will unfold in time, an unfolding that can be traced through the different astrological techniques of transits, secondary progressions and solar directions. Since the planets move in a rather orderly and predictable fashion, the temptation is then great to usurp God's role and *predict*. However, while one can anticipate the symbolic manifestation of an archetype, predicting its concrete actualisation is a much more ambitious and fraught endeavour. This is the pitfall into which fell many astrologers – attracting the wrath of Augustine, Ficino and Pico. Modern practitioners of astrology tend to steer clear of actual predictions – the days of 'meeting the tall dark stranger' are numbered – but the temptation to don the mantle of the omniscient seer is sometimes hard to resist.

Working under the premise of an objective quality of time, Jung, torn as ever between the mystic and the scientist, set to try and prove a law of synchronicity[2] operating between an objective psyche and objective events. This is the famed astrological experiment,[3] in which the psychologist, using astrological charts of married couples, tried to prove a prevalence of specific astrological contacts symbolising 'marriage' – namely, Sun conjunct Moon, Moon conjunct Moon and/or Ascendant conjunct Sun. In *Jung and Astrology*,[4] Maggie Hyde has made an excellent investigation into Jung's work and has dubbed Jung's initial 'objective' approach as Synchronicity I – an approach that did not yield any result:

> From his Synchronicity I perspective, Jung makes the mistake of confusing a piece of astronomical data with a symbol, so that astronomy is matched up with character traits and the subjective interpretation of the astrologer who derives meaning from the symbol is not part of the equation.[5]

Jung was drawing a blank with his experiment, but at this point things became unexpectedly weird and interesting. As he was sitting outside in his garden, pondering his lack of results, he suddenly saw "… a mischievous face laughing at him from the masonry of the wall… The thought struck him: had Mercurius, the spirit of nature, played a trick on him?"[6] Mercurius of course is none other than Hermes – and thus are we entering the hermeneutic loop. Jung proceeded to repeat the experiment, but this time he asked three people "whose psychological status was accurately known"[7] to carry it out in his place. One of them was a woman "with strong inner opposition whose union and reconciliation constituted her main problem."[8] Her results highlighted a predominance of Sun/ Moon contacts among married couples: "the classical *coniunctio Solis et Lunae* as the symbol of the union of the opposites is clearly emphasized,"[9] Jung duly noted, concluding: "the psychic and physical event (namely, the subject's problems and choice of horoscope) correspond, it would seem, to the nature of the archetype in the background and could therefore represent a synchronistic phenomenon."[10] Jung realized that not only was the psyche of the observer not detached from the experiment, it was actually fully involved in it. This, he named "the secret mutual connivance" – the unavoidable collusion of the experimenter with the experience. This is Synchronicity II, as Hyde called it. By 1955, Jung had totally changed his view on an objective quality of time independent of the observer and was inclined to think of time as 'hypothetical time', with "its hints at imaginal possibility":[11] the hermeneutic loop, in all its richness. The secret mutual connivance has serious implications for the practitioners of astrology. This has been highlighted and expounded by Geoffrey Cornelius' work on the hermeneutics of divination. In *The Moment of Astrology* Cornelius strongly challenges the objectivity of the Seed Moment, and argues for a renewed divinatory approach to the ancient art of astrology – away from objective, if not mechanistic, interpretations. An approach in which astrologers recognise their own involvement in the astrological moment. *A te et a scientia*, as Cornelius says – 'by you and by science'. This debate is far from settled in astrological circles. In the main, most astrologers do not pay much attention to the hermeneutic of their practice. They do not need to: one can be an excellent astrologer without reflecting on the philosophical tenets of the art. When pushed, the majority of practitioners will probably espouse a rather objective view: the Heavens speak and their role is to interpret them as neutrally as possible. For what it's worth, I think that Cornelius makes a very strong case. I often feel that the interpretation says as much about the astrologer as about the client. How can one take oneself out of the way anyway? Filters we all are, and if the work is to be as clear a filter as possible, I feel that one should never lose sight of the mischievous face grinning at the centre of the chart.

Endnotes

Introduction

1. See in particular Catherine Thimmesh's *Team Moon* (2015). New York: HMH Publishing.
2. *Myths To Live By* (1972), J. Campbell. British edition, reprint 2000. London: Souvenir Press, p. 234.
3. *The Eternal Drama* (1994), E. Edinger. Boston: Shambhala Publications, p. 37.
4. Which begs the question: *where are the women in this story?*
5. *Greek Myths and Christian Mystery* (1963), H. Rahner. Reprint 1971. New York, NY: Biblo & Tannen, p. 191.
6. *The Symbolism of Evil* (1969), P. Ricoeur. Beacon Paperback, p. 351.
7. *The Super Natural* (2016), W. Strieber & J. Kripal. New York, NY: Jeremy P. Tarcher, p. 113.
8. Three thousand years after its first steps, a more than respectable age, astrology's ability to ruffle feathers is unsurpassed.
9. The main ones being natal, horary, electional and mundane.
10. I am indebted to Darby Costello, astrologer and friend, for pointing out this analogy.
11. *The Super Natural* (2016), W. Strieber & J. Kripal. New York, NY: Jeremy P. Tarcher, p. 116.
12. Ibid, pp. 114-115.
13. *Synchronicity: An Acausal Connecting Principle*, CW 8, CG Jung. See Appendix.
14. Luminaries: names given to the Sun and the Moon in the astrological chart.
15. Whitley Strieber, who wrote extensively about his encounters with otherworldly creatures, co-authored *The Super Natural* with J. Kripal. A fascinating dialogue if ever there was one.
16. *The Super Natural* (2016), W. Strieber & J. Kripal. New York,

NY: Jeremy P. Tarcher, p. 111.

17. The Chinese, for instance, who spectacularly landed on the dark side of the Moon in January 2019, gave their mission the name of a Chinese lunar goddess, Chang'e.

18. One may also note that NASA's stated mission to "return astronauts to the Moon by 2024, including the first woman and the next man" has been named after Artemis, Apollo's sister and Goddess of the Moon. See NASA website: https://www.nasa.gov/artemis/. Accessed 3 July 2019.

19. Selene was technically a Titan, not a Goddess – point taken.

20. On 25 May 1961, JFK made a speech before a joint session of Congress in which he famously declared: "I believe that this nation should commit itself to achieving the goal, before this decade is out, of landing a man on the Moon and returning him safely to the Earth."

21. NASA website. Available at https://history.nasa.gov/SP-4402/ch4.htm. Accessed 20 August 2018.

22. *The Moon: Symbol of Transformation* (2003), J. Cashford. 2016 Paperback edition. Carterton: The Greystones Press, p. 162.

23. https://www.gaiafoundation.org.

24. At this point, I have to spell out very clearly that I am not entertaining conspiracy theories in this work, although the very fact of their existence is not without interest.

25. Clarke, quoted in *First on the Moon* (1970), G. Farmer & DJ Hamblin. London: Michael Joseph Ltd. Epilogue by Arthur C. Clarke, p. 401.

26. Mitchell, quoted in *Earthrise* (2008), R. Poole. Yale University Press, p. 112.

27. *The Moon: Symbol of Transformation* (2003), J. Cashford. 2016 Paperback edition. Carterton: The Greystones Press, p. 191.

28. *Myths To Live By* (1972), J. Campbell. British edition, reprint 2000. London: Souvenir Press, p. 237.

29. In the Jungian sense: archetypes are "forms of instinct", "contents of the collective unconscious... present from the

beginning", "beneath the personal psyche". "The archetypes have, when they appear, a distinctly numinous character which can only be described as 'spiritual'."

30. The distance from Earth to Moon is 237,000 miles. In Kubrick's *The Shining*, we find the mysterious and evil Room 237. "There ain't *nothing* in Room 237," one of the characters in the movie forcefully expresses. This has been read by conspiracy theorists as an oblique statement by Kubrick acknowledging that the Moon landing was faked.

31. *The Wind Cries Mary* – Jimi Hendrix. From the *Are You Experienced* album (1967).

Chapter 1

1. *Androgyny* (1976), J. Singer. Reprint 2000. York Beach, ME: Nicolas-Hays, p. 132.

2. Saturn orbits the Sun in 29.45 years. At this age in life, Saturn returns therefore to the place it was occupying at birth. In keeping with Saturn's traditional attributes, it is often seen as a period of crisis, when the wheel is turning, sometimes with dramatic consequences.

3. 'Stars' in an astrological sense.

4. *Occidental Mythology* (1964), J. Campbell. Reprint 2011. London: Souvenir Press, p. 75.

5. Philip K. Dick, quoted in *The Super Natural* (2016), W. Strieber & J. Kripal. New York, NY: Jeremy P. Tarcher, p. 319.

6. In the *Timaeus*.

7. *The Book of the Sun*, quoted in *Marsilio Ficino* (2006), A. Voss. Berkeley, CA: North Atlantic Books, p. 190.

8. A recent study by the Universities of Edinburgh and Kent claims that our Palaeolithic forebears had extensive knowledge of astronomical phenomena, including the precession of the equinoxes. Some artworks from Lascaux are said to depict comets and other celestial happenings.

9. The oldest known sculpture is the so-called 'Lion-man of

Hohlenstein-Stadel Cave' and dates from 38,000 BCE. Its actual gender is still a subject of controversy.

10. *The Moon: Symbol of Transformation* (2003), J. Cashford. 2016 Paperback edition. Carterton: The Greystones Press, p. 170.
11. *Blood Relations: Menstruation and the Origins of Culture* (1991), C. Knight. Reprint Paperback 1995. Yale University Press, p. 39.
12. As an aside, the story, initially recorded by French author Charles Perrault, may have originated in an earlier work by Italian poet Giambattista Basile, a work interestingly entitled *Sun, Moon and Talia.*
13. *The Moon: Symbol of Transformation* (2003), J. Cashford. 2016 Paperback edition. Carterton: The Greystones Press, p. 172.
14. *The Goddesses and Gods of Old Europe* (1982), M. Gimbutas. University of California Press, p. 152.
15. "The identification of consciousness with the heroic-Apollonic mode forces one into the absurdities of Neumann... for whom consciousness is masculine, even in women" – *The Myth of Analysis* (1972), J. Hillman. Reprint 1998. Northwestern University Press, p. 289.
16. *Occidental Mythology* (1964), J. Campbell. Reprint 2011. London: Souvenir Press, p. 86.
17. *Pagan Meditations* (1986), G. Paris. Third Printing 2017. Thompson: Spring Publications, pp. 11-12.
18. Ibid, p. 218.
19. *The Fear of the Feminine* (1994), E. Neumann. Princeton: Princeton University Press, p. 114.
20. *The Moon: Symbol of Transformation* (2003), J. Cashford. 2016 Paperback edition. Carterton: The Greystones Press, p. 176.
21. *The Myth of the Goddess* (1993), A. Baring & J. Cashford. London: Penguin Books, p. 282.
22. Jane Harrison, quoted in *Occidental Mythology* (1964), J. Campbell. Reprint 2011. London: Souvenir Press, p. 24.
23. *The Greek Myths* (1955), R. Graves. Revised edition 1992.

London: Penguin Books, p. xxix.

24. Ibid.

25. *The White Goddess* (1961), R. Graves. Reprint 1967. Whitstable: Latimer Trend & Co, p. 10.

26. *Occidental Mythology* (1964), J. Campbell. Reprint 2011. London: Souvenir Press, p. 21.

27. Ibid, p. 162.

28. *The Mother: Archetypal Image in Fairy Tales* (1977), ML von Franz. Toronto: Inner City Books.

29. *The Odyssey*, xxii, 1, quoted in *The Wanderings of Ulysses* (1823), T. Taylor.

30. *The White Goddess* (1961), R. Graves. Reprint 1967. Whitstable: Latimer Trend & Co, p. 292.

31. Prof Gilbert Murray, quoted in *Occidental Mythology* (1964), J. Campbell. Reprint 2011. London: Souvenir Press, p. 163.

32. *The Origin of Consciousness in the Breakdown of the Bicameral Mind* (1976), J. Jaynes. Reprint 1990. Boston: Houghton Mifflin Company, p. 75.

33. *The Greeks and the Irrational* (1951), ER Dodds. University of California Press, p. 43.

34. *To Touch the Face of God* (2013), K. Oliver. Baltimore: Johns Hopkins University Press, p. 48.

35. *The Ravaged Bridegroom* (1990), M. Woodman. Toronto: Inner City Books, p. 19.

Chapter 2

1. *The White Goddess* (1961), R. Graves. Reprint 1967. Whitstable: Latimer Trend & Co, p. 10.

2. Ibid, p. 11.

3. *The Republic*, Plato, 517c.

4. "half man, half snake... [Typhon] was so large that his head often knocked against the stars" – *Occidental Mythology* (1964), J. Campbell. Reprint 2011. London: Souvenir Press, p. 22.

5. Ibid, p. 24.
6. Jane Harrison, quoted in *Occidental Mythology* (1964), J. Campbell. Reprint 2011. London: Souvenir Press, p. 25.
7. *Mankind and Mother Earth* (1976), A. Toynbee. Oxford: Oxford University Press, p. 265.
8. *The Moon: Symbol of Transformation* (2003), J. Cashford. 2016 Paperback edition. Carterton: The Greystones Press, p. 13.
9. *Aryan Sun-myths* (1889), S. Titcombe. Troy, NY: Nims and Knight, p. 49.
10. *Greek Myths and Christian Mystery* (1963), H. Rahner. Reprint 1971. New York, NY: Biblo & Tannen, p. 94.
11. In the preface to *Beyond Good and Evil*, Nietzsche disdainfully remarked: "Christianity is Platonism for the 'common people'."
12. *The Goddess in the Gospels* (1998), M. Starbird. Rochester: Bear & Co, p. 124.
13. *The Myth of the Goddess* (1993), A. Baring & J. Cashford. London: Penguin Books, p. 568.
14. *The Moon: Symbol of Transformation* (2003), J. Cashford. 2016 Paperback edition. Carterton: The Greystones Press, p. 257.
15. Ibid, p. 384.
16. Alan Watts, quoted in *The Myth of the Goddess* (1993), A. Baring & J. Cashford. London: Penguin Books, p. 566.
17. *The Moon: Symbol of Transformation* (2003), J. Cashford. 2016 Paperback edition. Carterton: The Greystones Press, p. 185.
18. In psychological astrology, it is acknowledged that a man is more likely to express his Sun in the first half of his life and his Moon in the second half – and vice-versa for a woman. See *The Astrologer, the Counsellor and the Priest* (1997) by J. Sharman-Burke & L. Greene. London: Centre for Psychological Astrology.
19. *Aryan Sun-myths* (1889), S. Titcombe. Troy, NY: Nims and Knight, p. 151.
20. *The Roman Cult of Mithras* (1990), M. Clauss. Reprint 2001.

New York, NY: Routledge, p. 64.

21. *Patterns in Comparative Religion* (1958), M. Eliade. Reprint 1996. First Bison Books, pp. 92-93.

22. *Thou Art That* (2001), J. Campbell. Novato, CA: New World Library, p. 89.

23. *Dionysos: Archetypal Image of Indestructible Life* (1976), C. Kerényi. Reprint 1996. Princeton: Princeton University Press, p. 52.

24. *In Search of Duende* (1975), F. Garcia Lorca. Reprint 2010. New Directions Books, p. 96.

25. *Art and Myth of the Ancient Maya* (2017), O. Chinchilla Mazariegos. Yale: Yale University Press, p. 164.

26. Ibid, p. 6.

27. An association that is further illustrated in Arcana number 2 of the Tarot deck – the High Priestess – the guardian of secret, arcane knowledge associated with the Moon.

28. *The Secret of the Incas* (1996), W. Sullivan. New York, NY: Three Rivers Press, p. 292.

29. Ibid, p. 299.

30. Ibid, p. 300.

31. Ibid.

32. *Marsilio Ficino* (2006), A. Voss. Berkeley, CA: North Atlantic Books, p. 11.

33. *The Astrological World of Jung's Liber Novus* (2018), L. Greene. Oxon: Routledge, p. 34.

34. *Entering the Mysteries* (2016), A. Versluis. Minneapolis, MN: New Cultures Press, p. 38.

35. *Artemis: Virgin Goddess of the Sun and Moon* (2005), S. d'Este. London: Avalonia, p. 81.

36. Ibid, p. 79.

37. *Psychology and Alchemy* (1968), CG Jung. Reprint 1993. London: Routledge, pp. 343-344.

38. *Eclipse of the Sun* (1990), J. McCrickard. Glastonbury: Gothic Image Publications, p. xix.

39. Ibid, p. 228.
40. Ibid, p. 228.
41. Ibid, p. 227.
42. *The City of the Sun* (1602), T. Campanella. Reprint 2017. Whithorn: Anodos Books, p. 40.
43. Ibid.
44. www.esotericarchives.com/dee/monad.htm. Accessed 9 October 2018.
45. *The Lunation Cycle* (1962), D. Rudhyar. Reprint 1982. Santa Fe: Aurora Press, p. 20.
46. Ibid, p. 23.
47. *The Pictorial Key to the Tarot* (1911), AE Waite. 2005 Edition. Mineola, NY: Dover Publications, p. 70.
48. Ibid, p. 72.
49. The Moon nodes are the points of intersection between the path of the Moon in her rotation around the Earth, and the path of the Sun, the ecliptic; they are therefore the points where Moon meets Sun. The Moon crosses the ecliptic twice a lunar month.
50. *The Book of the Sun*, quoted in *Marsilio Ficino* (2006), A. Voss. Berkeley, CA: North Atlantic Books, p. 190.
51. Ibid, p. 195.
52. *Marsilio Ficino* (2006), A. Voss. Berkeley, CA: North Atlantic Books, p. 14.
53. Ibid, p. 25.
54. *The Book of the Sun*, quoted in *Marsilio Ficino* (2006), A. Voss. Berkeley, CA: North Atlantic Books, p. 212.

Chapter 3

1. *Homeric Hymns* (2003), translated by J. Cashford. London: Penguin Books, p. 27.
2. Cylon of Athens, a former victor of the Olympic Games, was an actual Athenian noble of the 7th century BCE who attempted, without success, to seize the city in a coup.

3. Diana is the Roman equivalent of Artemis.
4. *The Nature of the Gods*, Cicero. 2008 Edition. Oxford: Oxford University Press, p. 72.
5. Leo is ruled by the Sun – a fitting astrological sign for a Sun-God.
6. By sacrificing his daughter Iphigenia so as to obtain favourable winds when departing for Troy, Agamemnon unleashed a relentless family curse that, eventually, found resolution with Orestes, pursued by the Furies but redeemed by the solar goddess Athena.
7. Asclepius was then placed among the stars as the constellation Ophiuchus, the 'snake bearer'. Placed between Scorpio and Sagittarius, it crosses the ecliptic and is therefore sometimes referred to as the 13th sign of the zodiac.
8. *The White Goddess* (1961), R. Graves. Reprint 1967. Whitstable: Latimer Trend & Co, p. 14.
9. *The Myth of the Goddess* (1993), A. Baring & J. Cashford. London: Penguin Books, p. 660.
10. *The Moon: Symbol of Transformation* (2003), J. Cashford. 2016 Paperback edition. Carterton: The Greystones Press, p. 160.
11. *In the Dark Places of Wisdom* (1999), P. Kingsley. Inverness, CA: The Golden Sufi Center, p. 87.
12. *The Greek Myths* (1955), R. Graves. Revised edition 1992. London: Penguin Books, p. 52.
13. "the arrow… may sometimes be assimilated to the solar ray" – *Symbols of Sacred Science* (1962), R. Guénon. English translation of second French edition 2001, first published 2004. Hillsdale, NY: Sophia Perennis, p. 173.
14. *In the Dark Places of Wisdom* (1999), P. Kingsley. Inverness, CA: The Golden Sufi Center, p. 133.
15. Ibid, p. 134.
16. *Ancient Greek Divination* (2008), S. Iles Johnston. Chichester: Wiley-Blackwell, p. 57.
17. *Dionysos: Archetypal Image of Indestructible Life* (1976), C.

Kerényi. Reprint 1996. Princeton: Princeton University Press, p. 207.

18. *Ancient Greek Divination* (2008), S. Iles Johnston. Chichester: Wiley-Blackwell, p. 57.

19. Ibid, p. 56.

20. *The Greeks and the Irrational* (1951), ER Dodds. University of California Press, p. 70.

21. *Phaedrus*, Plato. Translated by Benjamin Jowett. Pantianos Classics, p. 74.

22. https://pythiaofdelphi.weebly.com/pythia-prophecies.html. Accessed 15 October 2018.

23. *Battlestar Galactica*, Season 4, Episode 2.

24. *Soluble Fish* (1924), A. Breton. Found in *Manifestoes of Surrealism*. First edition as an Ann Arbor paperback 1972. University of Michigan Press, p. 99.

25. *Manifesto of Surrealism* (1924), A. Breton. Found in *Manifestoes of Surrealism*. First edition as an Ann Arbor paperback 1972. University of Michigan Press, p. 36.

26. *In Search of Duende* (1975), F. Garcia Lorca. Reprint 2010. New Directions Books, p. 59.

27. *The Birth of Tragedy* (1872), F. Nietzsche. Amazon Printing, p. 19.

28. Ibid, p. 16.

29. *Howl* (1957), A. Ginsberg. City Lights Publishers.

30. Ibid.

31. *Cosmos and Psyche* (2006), R. Tarnas. First Plume Printing May 2007. Plume, p. 166.

32. *The Birth of Tragedy* (1872), F. Nietzsche. Amazon Printing, p. 25.

33. *The Four Faces of the Universe* (2006), R. Kleinman. Twin Lakes, WI: Lotus Press, p. 115.

34. *Artemis: Virgin Goddess of the Sun and Moon* (2005), S. d'Este. London: Avalonia, p. 35.

35. *The White Goddess* (1961), R. Graves. Reprint 1967. Whitstable:

Latimer Trend & Co, p. 285.

36. Callimachus, quoted in *Homo Necans* (1972), W. Burkert. 1983 edition. University of California Press, p. 123.

37. *Homo Necans* (1972), W. Burkert. 1983 Edition. University of California Press, p. 124.

38. *Dionysos: Archetypal Image of Indestructible Life* (1976), C. Kerényi. Reprint 1996. Princeton: Princeton University Press, p. xxxv.

39. *Return of the Goddess* (1982), E. Whitmont. New York, NY: The Crossroad Publishing Company, p. 85.

40. *Woman, Earth and Spirit* (1981), H. Luke. New York, NY: The Crossroad Publishing Company, p. 97.

41. Ibid.

42. *Mortals and Immortals* (1991), JP Vernant. Princeton: Princeton University Press, p. 195.

43. Ibid, p. 196.

44. *Artemis: Virgin Goddess of the Sun and Moon* (2005), S. d'Este. London: Avalonia, p. 109.

45. Ibid, p. 75.

46. *In the Dark Places of Wisdom* (1999), P. Kingsley. Inverness, CA: The Golden Sufi Center, p. 112.

47. Ibid.

48. Ibid, p. 89.

49. Ibid, pp. 91-92.

50. For Marie-Louise von Franz, the vision of the Midnight Sun referred to initiation into the Isis mysteries. Isis was called "mistress of light in the realm of darkness". The "initiation is a descent to the underworld and an illumination by a principle of consciousness which comes from the unconscious", in which is held the vision of "the gods celestial and the gods infernal" – *The Golden Ass of Apuleius* (1970), ML von Franz. 1992 edition. Boston, MA: Shambhala Publications, pp. 221-222.

51. *Woman, Earth and Spirit* (1981), H. Luke. New York, NY: The

Crossroad Publishing Company, p. 98.

52. *Sexual Personae* (1991), C. Paglia. New York, NY: First Vintage Books, p. 12.

53. Ibid, p. 13.

54. *Pagan Meditations* (1986), G. Paris. Third printing 2017. Spring Publications, p. 196.

55. Ibid, p. 197.

56. *LIFE Magazine*, quoted in *Moonfire* (1970), N. Mailer. Taschen, p. 144.

57. *The Myth of Analysis* (1972), J. Hillman. Paperback edition 1998. Northwestern University Press, p. 250.

58. Ibid.

59. Ibid, p. 258.

Chapter 4

1. *The Copernican Revolution* (1957), T. Kuhn. Twenty-fourth printing 2003. Harvard, pp. 179-180.

2. *The Sidereal Messenger*, G. Galilei. London: Rivingtons, p. 9.

3. *The Human Condition* (1958), H. Arendt. Second edition 1998. The University of Chicago Press, p. 275.

4. Ibid, p. 279.

5. *The Mysterium Lectures* (1995), E. Edinger. Toronto: Inner City Books, p. 297.

6. *The Master and His Emissary* (2009), I. McGilchrist. 2012 edition. Yale University Press, p. 43.

7. Ibid, p. 41.

8. Ibid, p. 428.

9. *Patterns in Comparative Religion* (1958), M. Eliade. 1996 edition. First Bison Books, p. 173.

10. *The Inner Reaches of Outer Space* (1986), J. Campbell. First paperback printing 2012. Novato, CA: New World Library, p. 1.

11. *The Master and His Emissary* (2009), I. McGilchrist. 2012 edition. Yale University Press, p. 429.

12. Ibid, p. 450.
13. Ibid, p. 93.
14. *The God Delusion* (2006), R. Dawkins. Reissued 2016. London: Transworld Publishers, p. 37.
15. *The Master and His Emissary* (2009), I. McGilchrist. 2012 edition. Yale University Press, p. 426.
16. *The God Delusion* (2006), R. Dawkins. Reissued 2016. London: Transworld Publishers, p. 34.
17. Ibid, p. 74.
18. *The Master and His Emissary* (2009), I. McGilchrist. 2012 edition. Yale University Press, p. 235.
19. *The God Delusion* (2006), R. Dawkins. Reissued 2016. London: Transworld Publishers, p. 72.
20. *The Human Condition* (1958), H. Arendt. Second edition 1998. The University of Chicago Press, p. 288.
21. *The Master and His Emissary* (2009), I. McGilchrist. 2012 edition. Yale University Press, p. 209.
22. http://www.independent.co.uk/voices/the-real-romance-in-the-stars-1527970.html. Accessed 29 October 2018.
23. *The God Delusion* (2006), R. Dawkins. Reissued 2016. London: Transworld Publishers, p. 56.
24. Ibid, pp. 115-116.
25. Ibid, p. 115.
26. *The Master and His Emissary* (2009), I. McGilchrist. 2012 edition. Yale University Press, p. 441.
27. Foreword by N. Cobb to *The Planets Within* (1982), T. Moore. Lindisfarne Books.
28. *Saving the Appearances* (1965), O. Barfield. Second edition 1988. Middletown, CT: Wesleyan University Press, p. 42.
29. Ibid, p. 14?
30. Hegel, quoted in *The Master and His Emissary* (2009), I. McGilchrist. 2012 edition. Yale University Press, p. 450.
31. *Saving the Appearances* (1965), O. Barfield. Second edition 1988. Middletown, CT: Wesleyan University Press, p. 143.

32. *The Human Condition* (1958), H. Arendt. Second edition 1998. The University of Chicago Press, p. 281.
33. *The Vocation Lectures* (1917), M. Weber. 2004 edition. Indianapolis, Indiana: Hackett Publishing Company, Inc., p. 30.
34. Ibid, pp. 12-13.
35. "Enchantment and modernity" (2012) in *PAN* no. 9, P. Curry, p. 81.
36. Ibid, p. 77.
37. *The God Delusion* (2006), R. Dawkins. Reissued 2016. London: Transworld Publishers, pp. 31-32.
38. Ibid.
39. Ibid.
40. At which point I might mention that his Sun is placed in Pisces and that therefore God might be lurking in the stars. But here is not the point.
41. *Saving the Appearances* (1965), O. Barfield. Second edition 1988. Middletown, CT: Wesleyan University Press, p. 143.
42. *The Master and His Emissary* (2009), I. McGilchrist. 2012 edition. Yale University Press, p. 433.
43. *The Vocation Lectures* (1917), M. Weber. 2004 edition. Indianapolis, Indiana: Hackett Publishing Company, Inc., p. 13.
44. *Science and the Modern World* (1925), AN Whitehead. First paperback edition 1967. New York, NY: The Free Press, p. 17.
45. "Iron Cage, Iron Crown" (2005), P. Curry, p. 6.
46. *After Prophecy: Imagination, Incarnation, and the Unity of the Prophetic Tradition* (2007), T. Cheetham. New Orleans, Louisiana: Spring Journal, Inc., pp. 161-162.
47. "The Overview Effect is a cognitive shift in awareness reported by some astronauts… during space flight." See chapter 7 for further elucidation.
48. *The Overview Effect* (1987), F. White. Third edition 2014.

Reston, Virginia: The American Institute of Aeronautics and Astronautics, p. xx.

49. Ibid, p. xxi.

50. NASA's avowed goal for the future *Artemis* missions "to lay the foundations for private companies to build a *lunar economy*" (my italics) flies in the face of any hope for a change of consciousness. Available at https://www.nasa.gov/artemis/. Accessed 3 July 2019.

51. *The Master and His Emissary* (2009), I. McGilchrist. 2012 edition. Yale University Press, p. 450.

52. Ibid, p. 49.

53. *The Fear of the Feminine* (1994), E. Neumann. Princeton: Princeton University Press, pp. 184-185.

Chapter 5

1. https://history.nasa.gov/SP-4402/ch4.htm. Accessed 2 November 2018.

2. The *Traveling Wilburys* were a rock supergroup formed in 1988 and composed of Bob Dylan, George Harrison, Jeff Lynne, Roy Orbison and Tom Petty.

3. *The Argonautica*, Apollonius of Rhodes. 2008 edition. Oxford: Oxford University Press, p. 16.

4. https://history.nasa.gov/SP-4402/ch4.htm. Accessed 2 November 2018.

5. *The Dimensions of Paradise* (1971), J. Michell. Second US edition. Rochester, Vermont: Inner Traditions, p. 55.

6. Quoted in *To Touch the Face of God* (2013), K. Oliver. Baltimore, Maryland: Johns Hopkins University Press, p. 39.

7. The design of the Mercury patch combines the astrological glyph of the planet Mercury with the number 7 – making it visually strikingly close to John Dee's *Monas Hieroglyphica*. See https://history.nasa.gov/project_mercury.jpg.

8. Ibid.

9. In *The Overview Effect* (1987), author Frank White stresses

that "almost from the beginning of his presidency, Kennedy repeatedly reached out to the Soviet leaders, proposing that the two superpowers work together, rather than competing on the space frontier" (p. 31).

10. One may note that in 2001 NASA referred back to the Argonauts by launching a satellite named *Jason-1*, whose remit was ocean mapping.

11. *The Overview Effect* (1987), F. White. Third edition 2014. Reston, Virginia: The American Institute of Aeronautics and Astronautics, p. 16.

12. *The Red Book, A Reader's Edition* (2009), CG Jung. First edition. New York, NY: Norton & Company, p. 314.

13. *The Eternal Drama* (1994), E. Edinger. Boston, MA: Shambhala Publications, Inc., p. 70.

14. Jungian Barbara Black Koltuv highlights four qualities of the rejected dark Goddess: 1. her lunar consciousness, her connections to the cycles of death and rebirth; 2. her body, instinctuality and sexuality; 3. her prophetic inner knowledge and experience, above logic or law; 4. her nature as Mother and Creatrix, an equal consort of God – *The Book of Lilith* (1986), BB Koltuv. Lake Worth, FL: Nicolas-Hays, Inc., pp. 121-122.

15. See next chapter.

16. *The Argonautica*, Apollonius of Rhodes. 2008 edition. Oxford: Oxford University Press, p. 11.

17. Ibid, p. 14.

18. *The Eternal Drama* (1994), E. Edinger. Boston, MA: Shambhala Publications, Inc., p. 65.

19. *The Argonautica*, Apollonius of Rhodes. 2008 edition. Oxford: Oxford University Press, p. 17.

20. Ibid.

21. Ibid, pp. 17-18.

22. *The Eternal Drama* (1994), E. Edinger. Boston, MA: Shambhala Publications, Inc., p. 66.

23. *The Odyssey*, ix, 94, quoted in *The Wanderings of Ulysses* (1999), T. Taylor.
24. *The Eternal Drama* (1994), E. Edinger. Boston, MA: Shambhala Publications, Inc., p. 68.
25. *The Argonautica*, Apollonius of Rhodes. 2008 edition. Oxford: Oxford University Press, p. 95.
26. *The Eternal Drama* (1994), E. Edinger. Boston, MA: Shambhala Publications, Inc., p. 67.
27. *The Argonautica*, Apollonius of Rhodes. 2008 edition. Oxford: Oxford University Press, p. 102.
28. *The Myth of the Goddess* (1993), A. Baring & J. Cashford. London: Penguin Books, p. 294.
29. *The Origins and History of Consciousness* (1954), E. Neumann. 2014 edition. Princeton: Princeton University Press, p. 161.
30. *The Ravaged Bridegroom* (1990), M. Woodman. Toronto: Inner City Books, pp. 18-19.
31. Ibid, p. 18.
32. *The Maiden King* (1998), R. Bly & M. Woodman. New York, NY: Henry Holt & Co, pp. 137-138.
33. *Anima* (1985), J. Hillman. Eighth Printing 2008. New York, NY: Spring Publications, p. 93.
34. Ibid.
35. Ibid, p. 95.
36. Ibid.
37. Ibid, p. 117.
38. *The Waste Land* (1922), TS Eliot. Reprint 2003. London: Penguin Classics.
39. *The Argonautica*, Apollonius of Rhodes. 2008 edition. Oxford: Oxford University Press, p. 102.
40. Ibid.
41. Ibid.
42. *The Eternal Drama* (1994), E. Edinger. Boston, MA: Shambhala Publications, Inc., p. 68.
43. *The Death of Nature* (1980), C. Merchant. First paperback

edition 1989. New York, NY: Harper Collins, p. 189.

44. Ibid, p. xvi.

45. Ibid, p. 28.

46. *I and Thou* (1970), M. Buber. 1996 edition. New York, NY: Touchstone, p. 53.

47. Ibid, p. 62.

48. "The *Minne* poetry... had as its subject matter the emotions, the sorrows and joys of the lover, and love itself, whether as a simple human emotion or as a mystical experience" – Jung & von Franz.

49. *The Grail Legend* (1960), E. Jung & ML von Franz. 1998 edition. Princeton University Press, p. 218.

Chapter 6

1. Aries, Leo and Sagittarius.

2. Neil's Moon was in Sagittarius though.

3. *First Man: The Life of Neil A. Armstrong* (2005), J. Hansen. 2012 edition. London: Simon & Schuster, p. 63.

4. *Moonfire* (1970), N. Mailer. Taschen, p. 58.

5. *First on the Moon* (1970), G. Farmer & DJ Hamblin. London: Michael Joseph Ltd, p. 159.

6. Ibid.

7. Ibid.

8. *Magnificent Desolation* (2009), B. Aldrin. Paperback 2010 edition. London: Bloomsbury Publishing, p. 34.

9. Ibid, p. 60.

10. Ibid, p. 61.

11. *First on the Moon* (1970), G. Farmer & DJ Hamblin. London: Michael Joseph Ltd, pp. 158-159.

12. Ibid, p. 290.

13. Ibid, p. 141.

14. *Aion* (1959), CG Jung. Second edition, Fifth printing 1981. London: Routledge, p. ix.

15. Ibid.

16. *Jung and Astrology* (1992), M. Hyde. London: The Aquarian Press, p. 19.
17. *Aion* (1959), CG Jung. Second edition, Fifth printing 1981. London: Routledge, p. 41.
18. Ibid, p. 77.
19. *Jung and Astrology* (1992), M. Hyde. London: The Aquarian Press, p. 21.
20. *The Last Man on the Moon* (1999), E. Cernan. New York, NY: St Martin's Press, p. 19.
21. For a deeper exploration of the hermeneutics of divination, I refer the reader to Geoffrey Cornelius' magisterial *The Moment of Astrology* (2003). Bournemouth: The Wessex Astrologer.
22. *Fixed Stars* (1998), B. Brady. York Beach, ME: Samuel Weiser, Inc., p. 171.
23. Available at https://history.nasa.gov/apollo.jpg.
24. *Fixed Stars* (1998), B. Brady. York Beach, ME: Samuel Weiser, Inc., p. 165.
25. Ibid, p. 169.
26. One may also note that the spacecraft used for the future *Artemis* missions to the Moon and beyond will be named *Orion*.
27. Orion is also a prime masonic symbol.
28. *Greek Myths* (1955), R. Graves. 1992 edition. London: Penguin, p. 133.
29. Each 'Sun age' covers a period of 5,125 years. The fifth Sun was therefore born on 21 December 2012 – the date of the famed Mayan prophecy.
30. One may also note that singer Sting, born on a New Moon, had his Sun positioned at 8 Libra – that is, aligned with NASA's Sun and with the Moon of the Moon landing chart. One of Sting's major successes, of course, was *Walking on the Moon*. The song was released on 4 November 1979, on a Full Moon.

31. Electional astrology: a branch of astrology that is used to choose a suitable time for big events (e.g. historically, the founding of a city or the coronation of a King/Queen).

32. That is, a range of +/- 1 degree between the planets: with NASA's Sun at 8 Libra, the Moon of the landing would have to be between 7 and 9 degrees Libra to be conjunct.

33. https://www.nasa.gov/mission_pages/apollo/missions/ apollo11.html. Accessed 26 November 2018.

34. *First on the Moon* (1970), G. Farmer & DJ Hamblin. London: Michael Joseph Ltd, p. 195.

35. Let us add the extraordinary fact that when JFK committed on 25 May 1961 to land a man on the Moon by the end of the decade, the Moon was also astrologically placed between 7 and 9 degrees Libra (assuming that the speech was made between 11am and 4pm).

36. See Appendix, *Astrology: When the Prince Kissed the Frog*.

37. *First on the Moon* (1970), G. Farmer & DJ Hamblin. London: Michael Joseph Ltd, p. 228.

38. Called a stellium.

39. *First on the Moon* (1970), G. Farmer & DJ Hamblin. London: Michael Joseph Ltd, p. 228.

40. Ibid.

41. Ibid.

42. *Zen Buddhism and Psychoanalysis* (1970), Suzuki, Fromm & De Martino. Revised edition. New York, NY: First Harper Colophon, pp. 5-6.

43. *Une histoire symbolique du Moyen Age Occidental* (2004), M. Pastoureau. Paris: Editions du Seuil, pp. 57-58.

44. Ibid, p. 85.

45. The solar return is the horoscope cast for the moment when the Sun returns at its natal position every year. It is the chart of one's birthday, and it gives a flavour of the year to come.

46. Available at https://history.nasa.gov/patches/Apollo/Apollo 11.jpg.

47. *First on the Moon* (1970), G. Farmer & DJ Hamblin. London: Michael Joseph Ltd, p. 208.

48. Ibid.

49. Ibid.

50. In 1969, the Vatican officially acknowledged that the depiction of Mary Magdalene in the Gospel of Luke as a sinful and penitent prostitute was wrongful. In 2016, Pope Francis instated 22 July as Magdalene's day.

51. *First on the Moon* (1970), G. Farmer & DJ Hamblin. London: Michael Joseph Ltd, p. 209.

52. https://www.nasa.gov/feature/the-making-of-the-apollo-11-mission-patch. Accessed 3 February 2018.

53. *First on the Moon* (1970), G. Farmer & DJ Hamblin. London: Michael Joseph Ltd, p. 209.

54. Ibid.

55. Ibid.

56. Ibid.

57. In a poignant recognition of President Kennedy's key role in initiating the Moon programme, an anonymous note was left on JFK's grave on the day of the landing with the following words: "Mr. President, the *Eagle* has landed."

58. *The Masks of God: Occidental Mythology* (2001), J. Campbell. Reprint 2011. London: Souvenir Press, p. 181.

59. https://www.brainyquote.com/authors/neil-armstrong-quotes. Accessed 22 August 2017.

60. *The Ravaged Bridegroom* (1990), M. Woodman. Toronto: Inner City Books, p. 18.

61. https://www.brainyquote.com/search_results?q=buzz+aldrin. Accessed 23 August 2017.

62. https://humanityplus.org/philosophy/transhumanist-declaration/. Accessed 28 November 2018.

63. M. Eliade, quoted in *First on the Moon* (1970), G. Farmer & DJ Hamblin. London: Michael Joseph Ltd, p. 224.

64. *Myths, Dreams and Mysteries* (1957), M. Eliade. 1967 edition.

New York, NY: Harper Torchbooks, p. 66.

65. *First on the Moon* (1970), G. Farmer & DJ Hamblin. London: Michael Joseph Ltd, p. 224.

Chapter 7

1. https://www.dailymotion.com/video/x1y5vf. Accessed 9 January 2019.

2. *Patterns in Comparative Religion* (1958), M. Eliade. 1996 edition. First Bison Books, pp. 125-126.

3. *The Inner Reaches of Outer Space* (1986), J. Campbell. First paperback printing 2012. Novato, CA: New World Library, pp. xx-xxi.

4. See chapter 4.

5. W. Schirra, quoted in *To Touch the Face of God* (2013), K. Oliver. Baltimore, Maryland: Johns Hopkins University Press, p. 96.

6. *The Thought of the Heart and the Soul of the World* (1992), J. Hillman. 2014 edition. New York, NY: Spring Publications, p. 36.

7. Ibid, p. 37.

8. S. Carpenter, quoted in *The Overview Effect* (2014), F. White. Third edition 2014. Reston, Virginia: The American Institute of Aeronautics and Astronautics, p. 29.

9. C. Kraft, quoted in *The Overview Effect* (2014), F. White. Third edition 2014. Reston, Virginia: The American Institute of Aeronautics and Astronautics, p. 29.

10. R. Chaffee, quoted in *The Overview Effect* (2014), F. White. Third edition 2014. Reston, Virginia: The American Institute of Aeronautics and Astronautics, p. 29.

11. Armstrong, quoted in *First on the Moon* (1970), G. Farmer & DJ Hamblin. London: Michael Joseph Ltd, p. 281.

12. *First on the Moon* (1970), G. Farmer & DJ Hamblin. London: Michael Joseph Ltd, p. 163.

13. M. Collins, quoted in *To Touch the Face of God* (2013), K.

Oliver. Baltimore, Maryland: Johns Hopkins University Press, p. 105.

14. *To Touch the Face of God* (2013), K. Oliver. Baltimore, Maryland: Johns Hopkins University Press, p. 105.

15. Armstrong, quoted in *First on the Moon* (1970), G. Farmer & DJ Hamblin. London: Michael Joseph Ltd, p. 296.

16. *Magnificent Desolation* (2009), B. Aldrin. Paperback 2010 edition. London: Bloomsbury Publishing, p. 38.

17. M. Collins, quoted in *The Overview Effect* (2014), F. White. Third edition 2014. Reston, Virginia: The American Institute of Aeronautics and Astronautics, p. 37.

18. A. Bean, ibid, p. 17.

19. It is striking how martial the vocabulary used in mountaineering is – assault, retreat, conquest, vanquish, defeat. A very solar, heroic consciousness.

20. *The Alpine Journal* (1866), E. Whymper, p. 151.

21. *Magnificent Desolation* (2009), B. Aldrin. Paperback 2010 edition. London: Bloomsbury Publishing, p. 35.

22. I am referring to the easiest route to the summit, via the Hörnli ridge. This route is equipped in places with fixed ropes, facilitating what would otherwise be a much more difficult climb.

23. B. Aldrin, quoted in *Earthrise* (2008), R. Poole. Yale University Press, p. 112.

24. G. Cernan, ibid.

25. M. Collins, quoted in *Earthrise* (2008), R. Poole. Yale University Press, p. 191.

26. https://www.nasa.gov/exploration/home/19jul_seaoftra nquillity.html. Accessed on 12 January 2019.

27. Nachman, quoted in *The Dream of the Cosmos* (2013), A. Baring. Dorset: Archive Publishing, p. 37.

28. C. Walker, quoted in *The Overview Effect* (2014), F. White. Third edition 2014. Reston, Virginia: The American Institute of Aeronautics and Astronautics, p. 21.

29. Available at http://www.chdr.cah.ucf.edu/spaceandspirit uality/files/2013/textualanalysis.pdf.

30. *The Overview Effect* (2014), F. White. Third edition 2014. Reston, Virginia: The American Institute of Aeronautics and Astronautics, p. 2.

31. M. Collins, quoted in *The Overview Effect* (2014), F. White. Third edition 2014. Reston, Virginia: The American Institute of Aeronautics and Astronautics, p. 183.

32. *The Overview Effect* (2014), F. White. Third edition 2014. Reston, Virginia: The American Institute of Aeronautics and Astronautics, p. 24.

33. Available at https://www.nasa.gov/image-feature/apollo-17-blue-marble.

34. Available at https://www.jpl.nasa.gov/spaceimages/details. php?id=PIA00452.

35. *Pale Blue Dot* (1994), C. Sagan. 1997 edition. Ballantine Books, p. 5.

36. Ibid, pp. 6-7.

37. Prince Sultan, quoted in *The Overview Effect* (2014), F. White. Third edition 2014. Reston, Virginia: The American Institute of Aeronautics and Astronautics, p. 241.

38. *The Master and His Emissary* (2009), I. McGilchrist. 2012 edition. Yale University Press, p. 300.

39. Ibid.

40. M. Collins, quoted in *The Overview Effect* (2014), F. White. Third edition 2014. Reston, Virginia: The American Institute of Aeronautics and Astronautics, p. 183.

41. Ibid, p. 10.

42. https://www.space-explorers.org. Accessed on 15 January 2019.

43. *The Way of the Explorer* (2016), E. Mitchell. New Page Books, p. 16.

44. *The Varieties of Religious Experience*, W. James. 1985 edition. London: Penguin Classics, pp. 380-381.

45. *The Shape of Light*, Suhrawardi. 2006 edition. Louisville, KY: Fons Vitae, p. 119.
46. *Cleansing the Doors of Perception* (2003), H. Smith. Boulder, CO: Sentient Publications, p. 11.
47. *The Overview Effect* (2014), F. White. Third edition 2014. Reston, Virginia: The American Institute of Aeronautics and Astronautics, p. 17.
48. For scholar Wouter Hanegraaf, "entheogens generate, or bring about, unusual states of consciousness in which those who use them are believed to be 'filled', 'possessed' or 'inspired' by some kind of divine entity, presence or force".
49. Gordon Wasson proposed that the psychoactive *kykeon* used in the Eleusinian Mysteries was made with barley parasitized by ergot.
50. Gordon Wasson, quoted in *Cleansing the Doors of Perception* (2003), H. Smith, p. 27.
51. *The Tao of Physics* (1975), F. Capra. 1991 edition. London: Flamingo, p. 12.
52. https://noetic.org/about/overview. Accessed 19 January 2019.
53. *Cleansing the Doors of Perception* (2003), H. Smith. Boulder, CO: Sentient Publications, p. 22.
54. Ibid, p. 23.
55. The first experience occurred under the influence of drugs, the second one did not.

Chapter 8

1. *Goddesses: Mysteries of the Feminine Divine* (2013), J. Campbell. Novato, CA: New World Library, p. 251.
2. *Woman's Mysteries* (1955), E. Harding. 1989 edition. London: Century, p. 168.
3. https://www.theoi.com/Khthonios/Hekate.html. Accessed 22 January 2019.
4. *Woman's Mysteries* (1955), E. Harding. 1989 edition. London:

Century, p. 26.

5. *The Moon: Symbol of Transformation* (2003), J. Cashford. 2016 Paperback edition. Carterton: The Greystones Press, p. 140.

6. Ibid.

7. Ibid, p. 182.

8. *Hekate Soteira* (1990), SI Johnston. The American Philological Association, p. 74.

9. *Patterns in Comparative Religion* (1958), M. Eliade. 1996 edition. First Bison Books, p. 183.

10. *The Moon: Symbol of Transformation* (2003), J. Cashford. 2016 Paperback edition. Carterton: The Greystones Press, p. 372.

11. *On The Face Which Appears On The Orb Of The Moon*, Plutarch. 1911 edition. London: Simpkin & Co, p. 27.

12. *The Lunation Cycle* (1967), D. Rudhyar. 1982 edition. Santa Fe, NM: Aurora Press, pp. 24-25.

13. *Greek Myths and Christian Mystery* (1963), H. Rahner. Reprint 1971. New York, NY: Biblo & Tannen, p. 160.

14. Ibid, p. 159.

15. Ibid, p. 160.

16. *The Divine Comedy*, Paradiso, Canto II.

17. *On The Face Which Appears On The Orb Of The Moon*, Plutarch. 1911 edition. London: Simpkin & Co, p. 48.

18. Ibid, p. 47.

19. Ibid, p. 45.

20. Ibid.

21. *The Moon: Symbol of Transformation* (2003), J. Cashford. 2016 Paperback edition. Carterton: The Greystones Press, p. 368.

22. Ibid, p. 379.

23. Pico, quoted in *Mysterium Coniunctionis* (1963), CG Jung. Second edition, seventh printing 1989. Princeton University Press, p. 144.

24. *Mysterium Coniunctionis* (1963), CG Jung. Second edition, seventh printing 1989. Princeton University Press, p. 146.

25. *The Lunation Cycle* (1982), D. Rudhyar. 1982 edition. Santa

Fe, NM: Aurora Press, p. 23.

26. *Hekate Soteira* (1990), SI Johnston. The American Philological Association, p. 1.
27. *Timaeus*, Plato, 36e.
28. *Hekate Soteira* (1990), SI Johnston. The American Philological Association, p. 49.
29. Ibid, pp. 3-4.
30. *De Mysteriis*, Iamblichus. Book 1.5.
31. *Hekate Soteira* (1990), SI Johnston. The American Philological Association, p. 34.
32. Ibid, p. 158.
33. *The Myth of Analysis* (1972), J. Hillman. Northwestern University Press, p. 51.
34. Ibid, pp. 51-52.
35. *Anima* (1985), J. Hillman. Eighth printing 2007. New York, NY: Spring Publications, p. 69.
36. Ibid, p. 52.
37. *The Thought of the Heart and the Soul of the World* (1992), J. Hillman. 2014 edition. New York, NY: Spring Publications, p. 62.
38. Ibid, p. 67.
39. Ibid, p. 69.
40. Ibid, p. 76.
41. "One must kill the adjective."
42. *The Thought of the Heart and the Soul of the World* (1992), J. Hillman. 2014 edition. New York, NY: Spring Publications, p. 79.
43. Ibid, p. 71.
44. *Swedenborg and Esoteric Islam* (1995), H. Corbin. Sixth printing 2014. West Chester, Pennsylvania: Swedenborg Foundation, p. 8.
45. *Alone with the Alone* (1969), H. Corbin. 1997 edition. Princeton University Press, p. 4.
46. "*Imaginatio* must not be confused with *fantasy*. As Paracelsus

observed, fantasy, unlike Imagination, is an exercise of thought without foundation in nature" – H. Corbin.

47. *Spiritual Body and Celestial Earth* (1989), H. Corbin. Fifth printing. Princeton University Press, p. 1.
48. *The World Turned Inside Out* (2015), T. Cheetham. New Orleans, Louisiana: Spring Journal Books, p. 112.
49. Association of Henry and Stella Corbin's Friends. www.amiscorbin.com.
50. Email received by the author, 12 December 2015 (author's translation).
51. *Alone with the Alone* (1969), H. Corbin. 1997 edition. Princeton University Press, p. 179.
52. Ibid, pp. 186-187.

Chapter 9

1. *First on the Moon* (1970), G. Farmer & DJ Hamblin. London: Michael Joseph Ltd, p. 228.
2. *Magnificent Desolation* (2009), B. Aldrin. Paperback 2010 edition. London: Bloomsbury Publishing, pp. 25-26.
3. Ibid, p. 26.
4. Woodruff, quoted in *First on the Moon* (1970), G. Farmer & DJ Hamblin. London: Michael Joseph Ltd, p. 229.
5. *Magnificent Desolation* (2009), B. Aldrin. Paperback 2010 edition. London: Bloomsbury Publishing, p. 27.
6. Ibid, p. 26.
7. Ibid, pp. 26-27.
8. https://history.nasa.gov/SP-368/s6ch1.htm. Accessed 30 January 2019.
9. *The Mysteries* (1955), CG Jung. Eranos Yearbooks, edited by Joseph Campbell. 1978 edition. Princeton University Press, Bollingen series, pp. 318-319.
10. Ibid, p. 318.
11. Ibid, p. 287.
12. *The Moon: Symbol of Transformation* (2003), J. Cashford. 2016

Paperback edition. Carterton: The Greystones Press, p. 184.

13. *Greek Myths and Christian Mystery* (1963), H. Rahner. Reprint 1971. New York, NY: Biblo & Tannen, p. 89.
14. Ibid, p. 122.
15. Ibid, p. 133.
16. Ibid, pp. 156-157.
17. *Thou Art That* (2001), J. Campbell. Novato, CA: New World Library, p. 113.
18. Ibid, pp. 104-105.
19. Ibid, p. 107.
20. *Greek Myths and Christian Mystery* (1963), H. Rahner. Reprint 1971. New York, NY: Biblo & Tannen, pp. 156-157.
21. Ibid, p. 167.
22. Ibid.
23. Ibid, p. 173.
24. Ibid, p. 174.
25. Ibid.
26. *Psychology and Alchemy* (1953), CG Jung. Second edition, reprinted 1993. London: Routledge, p. 312.
27. Ibid, p. 313.
28. Ibid.
29. *Mysterium Coniunctionis* (1963), CG Jung. Seventh printing 1989. Princeton University Press, p. xiv.
30. Ibid, p. 3.
31. Ibid, p. 89.
32. Ibid, p. 461.
33. *The Astrological World of Jung's Liber Novus* (2018), L. Greene. London: Routledge, p. 59.
34. Ibid.
35. Leo, quoted in *The Astrological World of Jung's Liber Novus* (2018), L. Greene. London: Routledge, p. 59.
36. *Mysterium Coniunctionis* (1963), CG Jung. Seventh printing 1989. Princeton University Press, p. 535.
37. *Psychology and Alchemy* (1953), CG Jung. Second edition,

reprinted 1993. London: Routledge, p. 345.

38. In the Hindu pantheon, a benevolent deity and a wife of Siva.

39. *Answer to Job* (1958), CG Jung. Second edition, 2010 reprint. Princeton University Press, Bollingen series, pp. 99-100.

40. Ibid, p. 100.

41. *Thou Art That* (2001), J. Campbell. Novato, CA: New World Library, p. 106.

42. *The Myth of the Goddess* (1993), A. Baring & J. Cashford. London: Penguin Books, p. 572.

43. *The Moon: Symbol of Transformation* (2003), J. Cashford. 2016 Paperback edition. Carterton: The Greystones Press, p. 180.

44. Ibid, p. 388.

45. Jung, quoted in *The Creation of Consciousness* (1984), E. Edinger. Toronto: Inner City Books, pp. 18-19.

46. *The Dimensions of Paradise* (1971), J. Michell. Second US edition, 2008. Rochester, VT: Inner Traditions, p. 19.

47. Ibid, p. 1.

48. See chapter 5.

49. *The Dimensions of Paradise* (1971), J. Michell. Second US edition, 2008. Rochester, VT: Inner Traditions, p. 7.

50. Ibid, p. 34.

51. Ibid.

52. Ibid, p. 206.

53. Ibid, p. 209.

54. *The Inner Reaches of Outer Space* (1986), J. Campbell. Reprint 2002. Novato, CA: New World Library, p. 11.

55. *Hamlet's Mill* (1969), G. de Santillana & H. von Dechend. Sixth printing, 2002. Jaffrey, New Hampshire: David R. Godine, p. 7.

56. *The Dimensions of Paradise* (1971), J. Michell. Second US edition, 2008. Rochester, VT: Inner Traditions, p. 213.

57. Ibid, p. 216.

58. Ibid, p. 223.

59. *The Law of Vibration* (2013), T. Plummer. Petersfield, Hampshire: Harriman House, p. 19.
60. Ibid.
61. *A Little Book of Coincidence in the Solar System* (2001), J. Martineau. Revised edition 2012. Glastonbury: Wooden Books Ltd.
62. *Sun, Moon and Earth* (1999), R. Heath. Revised edition 2012. Glastonbury: Wooden Books Ltd, p. 22.
63. Three times eleven is 33, a very important masonic number. Buzz Aldrin was in the 33rd Masonic degree when he set sail for the Moon, and he is said to have initiated his Communion ceremony 33 minutes after the *Eagle* touched down on the lunar surface.
64. *A Little Book of Coincidence in the Solar System* (2001), J. Martineau. Revised edition 2012. Glastonbury: Wooden Books Ltd, p. 30.
65. Ibid.
66. *The Dimensions of Paradise* (1971), J. Michell. Second US edition, 2008. Rochester, VT: Inner Traditions, pp. 117-118.
67. http://www.whats-your-sign.com/spiritualmeaning ofnumbereleven.html. Accessed 5 February 2019.
68. *The Dimensions of Paradise* (1971), J. Michell. Second US edition, 2008. Rochester, VT: Inner Traditions, p. 84.

Chapter 10

1. *The Thought of the Heart and the Soul of the World* (1992), J. Hillman. 2014 edition. New York, NY: Spring Publications, p. 52.

Chapter 11/1

1. Numen: a spiritual force or influence often identified with a natural object, phenomenon, or place (Merriam-Webster).
2. *The Idea of the Holy* (1923), R. Otto. Second edition 1950. Paperback 1958 print. Oxford University Press, p. 7.

3. Ibid, p. 13.
4. Ibid, p. 30.
5. *Psychology and Alchemy* (1953), CG Jung. Second edition, reprinted 1993. London: Routledge, p. 345.
6. *Gaia: A New Look at Life on Earth* (1979), J. Lovelock. Reissued 2000. Oxford University Press, p. 7.
7. Ibid, p. 6.
8. Ibid.
9. Ibid, p. 7.
10. Ibid.
11. Ibid, p. 10.
12. Ibid.
13. Ibid, p. ix.
14. Ibid, p. xi.
15. Intergovernmental Panel on Climate Change. https://www.ipcc.ch.
16. *Gaia: A New Look at Life on Earth* (1979), J. Lovelock. Reissued 2000. Oxford University Press, p. xiii.
17. Ibid, p. xv.
18. *The Fear of the Feminine* (1994), E. Neumann. Princeton University Press, pp. 168-169.
19. Ibid, p. 171.
20. *The Life Divine* (1949), Sri Aurobindo. Second American edition, fourth impression 2006. Twin Lakes, WI: Lotus Press, p. 9.
21. *The God Delusion* (2006), R. Dawkins. 2016 edition. London: Transworld Publishers, pp. 193-194.
22. *The Life Divine* (1949), Sri Aurobindo. Second American edition, fourth impression 2006. Twin Lakes, WI: Lotus Press, p. 9.
23. His constant use of the word 'primitive' to describe indigenous people is particularly shocking to our 21st century sensibility.
24. *The Notebooks on Primitive Mentality* (1949), L. Levy-Bruhl.

1978 edition. New York, NY: Harper & Row, p. 73.

25. Ibid, p. 71.
26. Ibid.
27. Ibid, p. 74.
28. *The Life Divine* (1949), Sri Aurobindo. Second American edition, fourth impression 2006. Twin Lakes, WI: Lotus Press, p. 15.
29. *A Feeling for the Organism* (1983), E. Fox Keller. 2003 edition. New York, NY: Holt Paperbacks, p. 204.
30. McClintock, quoted in *A Feeling for the Organism* (1983), E. Fox Keller. 2003 edition. New York, NY: Holt Paperbacks, p. 198.
31. Ibid, p. 204.
32. Ibid, p. 203.
33. *A Feeling for the Organism* (1983), E. Fox Keller. 2003 edition. New York, NY: Holt Paperbacks, p. 205.
34. McClintock, quoted in *A Feeling for the Organism* (1983), E. Fox Keller. 2003 edition. New York, NY: Holt Paperbacks, pp. 205-206.
35. "Woman [the Feminine] is the future of Mankind."
36. *The Myth of the Goddess* (1993), A. Baring & J. Cashford. London: Penguin Books, p. 304.
37. *Homeric Hymns* (2003). Translation by J. Cashford. London: Penguin.
38. One may note that in Aztec cosmology, the Earth was also the mother of the Sun and of the Moon.
39. *The Myth of the Goddess* (1991), A. Baring & J. Cashford. London: Penguin Books, p. 305.
40. See chapter 1.
41. *Myths To Live By* (1972), J. Campbell. British edition, reprint 2000. London: Souvenir Press, p. 244.
42. Ibid, p. 571.
43. Ibid, p. 572.
44. *The Life Divine* (1949), Sri Aurobindo. Second American

edition, fourth impression 2006. Twin Lakes, WI: Lotus Press, p. 10.
45. Ibid, p. 8.
46. *In Search of Duende* (1975), F. Garcia Lorca. Reprint 2010. New Directions Books, p. 59.
47. Ibid, p. 17.
48. *Thou Art That* (2001), J. Campbell. Novato, CA: New World Library, p. 36.
49. Ibid.
50. *Gaia*, written for the Gaia Foundation by Jules Cashford.
51. *Homeric Hymns* (2003). Translation by J. Cashford. London: Penguin.

Chapter 11/2
1. *The Waste Land* (1922), TS Eliot. 2002 edition. London: Faber and Faber.
2. *The Heresy of Self-Love* (1968), P. Zweig. 1980 edition. Princeton University Press, p. 92.
3. Ibid.
4. *The New View Over Atlantis* (1969), J. Michell. 1983 edition. London: Thames & Hudson, p. 8.
5. *The Life Divine* (2006), Sri Aurobindo. Second American edition, fourth impression 2006. Twin Lakes, WI: Lotus Press, p. 6.
6. McClintock, quoted in *A Feeling for the Organism* (1983), E. Fox Keller. 2003 edition. New York, NY: Holt Paperbacks, p. 200.
7. *The Philosophy of Civilization* (1949), A. Schweitzer. 1987 edition. Amherst, NY: Prometheus Books, p. 310.
8. Ibid, p. 317.
9. Ibid, pp. 317-318.
10. The ruling planet has a special role in the horoscope. It may be seen as the main soloist in the cosmological orchestra that is the astrological chart, the Sun being the conductor.

11. Esoteric astrology is an astrological system developed by theosophist Alice Bailey. See *Esoteric Astrology* (1951), A. Bailey. Lucis Publishing Companies.

12. *Evening Thoughts* (2006), T. Berry. San Francisco, CA: Sierra Club Books, p. 43.

13. Ibid, p. 18.

14. *Soulcraft* (2003), B. Plotkin. Novato, CA: New World Library, p. 237.

15. *The Philosophy of Civilization* (1949), A. Schweitzer. 1987 edition. Amherst, NY: Prometheus Books, p. 313.

16. *The Spell of the Sensuous* (1996), D. Abram. New York, NY: Vintage Books, p. 131.

17. *Spiritual Ecology* (2016), S. Murphy. Edited by Llewellyn Vaughan-Lee. The Golden Sufi Center, p. 116.

18. *Ecopsychology* (1995), P. Shepard. Edited by Roszak, Kanner & Gomes. Counterpoint, p. 35.

19. *The Spell of the Sensuous* (1996), D. Abram. New York, NY: Vintage Books, p. 22.

20. Ibid, p. 267.

21. Jung quoted in *The Earth Has a Soul: Jung on Nature, Technology and Modern Life* (2002), edited by M. Sabini. North Atlantic Books, p. 15.

22. *Dark Mountain*, Issue 2, Summer 2011, p. 71.

23. Ibid.

24. Joanna Macy is a leading voice in the environmental movement. Her work is deeply infused by Buddhist teachings. www.joannamacy.net.

25. Ibid.

26. *The Philosophy of Civilization* (1949), A. Schweitzer. 1987 edition. Amherst, NY: Prometheus Books, p. 318.

27. *The Spell of the Sensuous* (1996), D. Abram. New York, NY: Vintage Books, p. 72.

28. *Mindfulness and the Natural World* (2013), C. Thompson. Leaping Hare Press, p. 57.

Epilogue

1. *Tantra Illuminated* (2012), C. Wallis. Second edition 2013. Boulder, CO: Mattamayura Press, p. 186.
2. *Apollo 11: 50th anniversary of the first man on the Moon.* Published in 2019 by Mortons Group. ISBN 978-1-911276-82-1. Chapter 16, p. 127.
3. Ibid.

Appendix

1. "In Memory of Richard Wilhelm", in *The Secret of the Golden Flower* (1931), CG Jung. 2010 edition. San Diego, CA: The Book Tree, p. 143.
2. A rather elusive and ambiguous concept defined by Jung as "acausal meaningful coincidence". Roderick Main has contributed much to expand and clarify it in *Revelations of Chance* (2007). State University of New York Press.
3. *Synchronicity: An Acausal Connecting Principle* (1960), CG Jung. 2010 edition. Princeton University Press.
4. *Jung and Astrology* (1992), M. Hyde. London: The Aquarian Press. I would warmly encourage anyone with an interest in synchronicity and into the hermeneutic loop to read Maggie's book.
5. *Jung and Astrology* (1992), M. Hyde. London: The Aquarian Press, p. 130.
6. Von Franz, quoted in *Jung and Astrology* (1992), M. Hyde. London: The Aquarian Press, p. 130.
7. *Synchronicity: An Acausal Connecting Principle* (1960), CG Jung. 2010 edition. Princeton University Press, p. 57.
8. Ibid, p. 58.
9. Ibid.
10. Ibid, p. 59.
11. Ibid, p. 127.

Author's Biography

In a first incarnation, Philippe Sibaud was known as an oil trader. In 2006, an intense experience set him on a radically different path. He now co-runs Umunthu Microfinance, a small Malawian NGO that he set up in 2010 to provide small-scale loans to disadvantaged women in Malawi (currently 1,000 clients). He is a Trustee with the Gaia Foundation, an international London-based NGO which has been working with indigenous people for 35 years. In 2017, Philippe graduated from Canterbury Christ Church University with an MA in Myth, Cosmology and the Sacred (with Distinction). Over the course of twenty years, he has published four poetry books and one fiction novel (in French) and has been a dedicated student of astrology for many a solar return. He was born at night under a crisp February Full Moon.

Personal Message to the Reader

It has been a long journey. It has been an enriching journey, and it is far from over yet. The grass beckons. The stars impel. We are sparkles of Consciousness, each and every one of us. We reflect. We refract. We glow. We make mistakes. I do not make any claim for truth and I am willing to hear. If you so wish, please feel free to contact me at:

scavengersofbeauty@gmail.com

Blessings under the trees,

Philippe Sibaud

BOOKS

SPIRITUALITY

O is a symbol of the world, of oneness and unity; this eye
represents knowledge and insight. We publish titles on general
spirituality and living a spiritual life. We aim to inform and help
you on your own journey in this life.
If you have enjoyed this book, why not tell other readers by
posting a review on your preferred book site?

Recent bestsellers from O-Books are:

Heart of Tantric Sex
Diana Richardson
Revealing Eastern secrets of deep love and intimacy to Western couples.
Paperback: 978-1-90381-637-0 ebook: 978-1-84694-637-0

Crystal Prescriptions
The A-Z guide to over 1,200 symptoms and their healing crystals
Judy Hall
The first in the popular series of eight books, this handy little guide is packed as tight as a pill-bottle with crystal remedies for ailments.
Paperback: 978-1-90504-740-6 ebook: 978-1-84694-629-5

Take Me To Truth
Undoing the Ego
Nouk Sanchez, Tomas Vieira
The best-selling step-by-step book on shedding the Ego, using the teachings of *A Course In Miracles*.
Paperback: 978-1-84694-050-7 ebook: 978-1-84694-654-7

The 7 Myths about Love...Actually!
The Journey from your HEAD to the HEART of your SOUL
Mike George
Smashes all the myths about LOVE.
Paperback: 978-1-84694-288-4 ebook: 978-1-84694-682-0

The Holy Spirit's Interpretation of the New Testament
A Course in Understanding and Acceptance
Regina Dawn Akers
Following on from the strength of *A Course In Miracles*, NTI
teaches us how to experience the love and oneness of God.
Paperback: 978-1-84694-085-9 ebook: 978-1-78099-083-5

The Message of A Course In Miracles
A translation of the Text in plain language
Elizabeth A. Cronkhite
A translation of *A Course in Miracles* into plain, everyday
language for anyone seeking inner peace. The companion
volume, *Practicing A Course In Miracles*, offers practical lessons
and mentoring.
Paperback: 978-1-84694-319-5 ebook: 978-1-84694-642-4

Your Simple Path
Find Happiness in every step
Ian Tucker
A guide to helping us reconnect with what is really important in
our lives.
Paperback: 978-1-78279-349-6 ebook: 978-1-78279-348-9

365 Days of Wisdom
Daily Messages To Inspire You Through The Year
Dadi Janki
Daily messages which cool the mind, warm the heart and guide
you along your journey.
Paperback: 978-1-84694-863-3 ebook: 978-1-84694-864-0

Body of Wisdom
Women's Spiritual Power and How it Serves
Hilary Hart
Bringing together the dreams and experiences of women across
the world with today's most visionary spiritual teachers.
Paperback: 978-1-78099-696-7 ebook: 978-1-78099-695-0

Dying to Be Free
From Enforced Secrecy to Near Death to True Transformation
Hannah Robinson
After an unexpected accident and near-death experience, Hannah
Robinson found herself radically transforming her life, while a
remarkable new insight altered her relationship with her father, a
practising Catholic priest.
Paperback: 978-1-78535-254-6 ebook: 978-1-78535-255-3

The Ecology of the Soul
A Manual of Peace, Power and Personal Growth for Real People
in the Real World
Aidan Walker
Balance your own inner Ecology of the Soul to regain your
natural state of peace, power and wellbeing.
Paperback: 978-1-78279-850-7 ebook: 978-1-78279-849-1

Not I, Not other than I
The Life and Teachings of Russel Williams
Steve Taylor, Russel Williams
The miraculous life and inspiring teachings of one of the World's
greatest living Sages.
Paperback: 978-1-78279-729-6 ebook: 978-1-78279-728-9

On the Other Side of Love

A woman's unconventional journey towards wisdom
Muriel Maufroy
When life has lost all meaning, what do you do?
Paperback: 978-1-78535-281-2 ebook: 978-1-78535-282-9

Practicing A Course In Miracles

A translation of the Workbook in plain language, with mentor's notes
Elizabeth A. Cronkhite
The practical second and third volumes of The Plain-Language
A Course In Miracles.
Paperback: 978-1-84694-403-1 ebook: 978-1-78099-072-9

Quantum Bliss

The Quantum Mechanics of Happiness, Abundance, and Health
George S. Mentz
Quantum Bliss is the breakthrough summary of success and
spirituality secrets that customers have been waiting for.
Paperback: 978-1-78535-203-4 ebook: 978-1-78535-204-1

The Upside Down Mountain

Mags MacKean
A must-read for anyone weary of chasing success and happiness
– one woman's inspirational journey swapping the uphill slog for
the downhill slope.
Paperback: 978-1-78535-171-6 ebook: 978-1-78535-172-3

Your Personal Tuning Fork
The Endocrine System
Deborah Bates
Discover your body's health secret, the endocrine system, and
'twang' your way to sustainable health!
Paperback: 978-1-84694-503-8 ebook: 978-1-78099-697-4

Readers of ebooks can buy or view any of these bestsellers by
clicking on the live link in the title. Most titles are published
in paperback and as an ebook. Paperbacks are available in
traditional bookshops. Both print and ebook formats are
available online.

Find more titles and sign up to our readers' newsletter at
http://www.johnhuntpublishing.com/mind-body-spirit

Follow us on Facebook at https://www.facebook.com/OBooks/
and Twitter at https://twitter.com/obooks